George Huntington Williams

Elements of Crystallography

For Students of Chemistry, Physics and Mineralogy

George Huntington Williams

Elements of Crystallography
For Students of Chemistry, Physics and Mineralogy

ISBN/EAN: 9783337276188

Printed in Europe, USA, Canada, Australia, Japan

Cover: Foto ©berggeist007 / pixelio.de

More available books at **www.hansebooks.com**

ELEMENTS

OF

CRYSTALLOGRAPHY

FOR STUDENTS OF CHEMISTRY PHYSICS AND MINERALOGY

BY

GEORGE HUNTINGTON WILLIAMS PhD

ASSOCIATE PROFESSOR IN THE JOHNS HOPKINS UNIVERSITY

NEW YORK
HENRY HOLT AND COMPANY
1890

PREFACE.

The present book is the outgrowth of a long-felt personal need and the hope that a concise and elementary statement of the general principles of Crystallography may prove acceptable to other students than those of mineralogy.

To both chemists and physicists the subject is important, although the information in regard to it embraced in text-books and lectures on chemistry and physics is usually inadequate. Crystals are constantly employed as a means of purifying, recognizing and tracing the relationships between chemical compounds of all sorts; their genesis and growth offer interesting problems in molecular physics, and their completed forms furnish material for the study of elasticity, cohesion and the propagation of various forms of radiant energy.

To the geologist in every field, as well as to the mining engineer, Crystallography is the starting-point to a knowledge of mineralogy, while to the student of petrography an acquaintance with crystal form and its relation to the optical properties of crystals is indispensable.

Considerable experience has convinced the writer that Crystallography, in its simplest form, is well calculated to arouse the student's interest for its own sake, and that it may with advantage be incorporated into many courses where detailed instruction in mineralogy is impossible. This book is not intended as a complete treatise, but merely to furnish so much information on the subject as may be of service to students of other but allied branches. It is, however, hoped that it may also be of use as a skeleton for a more exhaustive presentation of the entire subject when such is desirable.

Such a purpose as this is sufficient excuse for the omission of much that is of cardinal importance to Crystallography as a whole. Mathematical treatment, the formulæ necessary for the calculation of constants and symbols from measured angles, the application of spherical projection, and all descriptions of the construction and manipulation of crystallographic instruments must be sought for in larger works, whose titles are given.

Those methods of presentation which experience has shown to be most readily grasped by the beginner have been throughout preferred to those which advanced workers may consider more elegant and satisfactory. The symbols of Weiss are taken as a starting-point, since they most clearly indicate the position of a plane with reference to the crystallographic axes. The shortened form of these symbols, suggested by Naumann, has been generally employed, although Miller's index symbols are written beside them, in order to familiarize the student simultaneously with both of these methods of notation. Naumann's symbols have been preferred to those of Dana, because of their more general use in works on crystallography and mineralogy; and because, to any one familiar with Naumann's, Dana's symbols can present no difficulty.

In subject-matter this little book makes no claim to originality. It is only an attempt to present to English students a clear and concise statement of the results secured by others. In plan and illustration, the admirable treatise of Groth (Physikalische Krystallographie, Leipzig, 1885) has been freely used. Suggestions and figures have also been taken from other sources whenever possible. My grateful acknowledgments are especially due to my friend Professor S. L. Penfield of New Haven, to whose generous advice and suggestion any value which this book may possess is in no small degree due.

BALTIMORE, July, 1890.

ERRATA.

p. 7, 5th line from bottom, after p insert . vii.

8, Fig. 6 wrong. Alternate revolved rows should be ⊘, not ⊘.

25 (bis), 49 (bis), 50, 86 (bis), 105 and 109, for *secant* read *sectant*

30, 3d line from top, for $1-\infty$ read $1-i$

32, 12th " " bottom, for $\infty a : a : a \quad \infty O \quad I \quad (011)$ read
$a : a : \infty a \; \infty O \quad I \quad (110)$

32, 6th " " " " $a : b : \infty c \; P\infty \; 1\text{-}i \; \{110\}$ read
$\infty a : b : c \quad P\infty \; 1\text{-}i \; \{011\}$

40, 14th line from top, for $\epsilon\delta\rho\alpha$ read $\tilde{\epsilon}\delta\rho\alpha$

43, top " for $\eta\mu\iota$ read $\acute{\eta}\mu\iota$-

56, 13th " from top, for i read I

64, 4th " " " " $\gamma\{hlk\}$ and $\gamma\{hkl\}$ read
$\gamma\{lkh\}$ and $\gamma\{klh\}$

78, transpose Figs. 109 and 110; also the statements below relative to genesis of pentagonal icositetrahedron and diploid.

80, 8th line from bottom, for *uralyl* read *uranyl*

86, 7th " " " " *possbile* " *possible*

91, 2d " " top, for $\{OP, 001\}$ " $OP, \{001\}$

95, 3d " " bottom, for (c), read (e),

97, last word, for *diacetylphenol phtalline* read *diacetylphenolphthaleïn*.

99, 4th line from top, for *helòhedrons* read *holohedrons*

100, 7th " " " " *toluosulphanide* read *toluenesulphonamide*

110, 9th line from bottom, for *three first* read *first three*

113, top line, for $\{\bar{h}kkl\}$ read $\{kk\bar{h}l\}$

113, 2d line from top, for $\{\bar{2}112\}$ read $\{11\bar{2}2\}$

119, 7th line from bottom, for $\{hi\bar{k}l\}$ read $\pi\{hikl\}$

128, Fig. 202, on lower right plane, for n read R

133, last line, for *trapezoedrons* read *trapezohedrons*

138, " " " $\pm \dfrac{mPn}{4}\cdot\dfrac{r}{l}$ read $\pm \dfrac{mP2}{4}\cdot\dfrac{l}{r}$

141, 8th line from top, for *sulpharsenide* read *sulpharsenite*

141, 9th " " " " *sulphantimonide* read *sulphantimonite*

145, Fig. 240, right lower angle of spherical triangle should reach only to the dotted line.

153, 5th line from bottom, for $[(Fe,Mg)SiO_3]$ read $[(Fe_2,Mg_2)SiO_4]$

157, 3d line from bottom, for *triphenylmethan* read *triphenylmethane*

157, 3d line from bottom, for $(C_6H_5)CH$ read $(C_6H_5)_3CH$

166, 4th " " top, for $P\infty$ read $P\widetilde{\infty}$

169, 4th " " " " *campheroxim* read *camphoroxime*

176, 13th " " bottom, for $\{100\}$ read $\{010\}$

185, top line, for *what* read *that*

186, for G. 289 read FIG. 289

205, Fig. 337, straighten.

208, 13th line from bottom, for *by* read *from*

223, 10th " " " first figure, for 1 read 0

223, 12th " " " last " for 1 " $\bar{1}$

227, 10th " " " for $\kappa\{31\bar{4}1\}$ read $\kappa\{13\bar{4}1\}$

228, for Fig. 368 read Fig. 363

240, 2d line from top, for $Od' = OD'.\sin\alpha$ read $Od' = Od.\sin\alpha$

BIBLIOGRAPHY.

The following list of references may be found of use by those desiring fuller information in regard to the subjects touched upon in this book:

I. On the Molecular Structure of Crystals.

R. J. HAUY: Essai d'une théorie de la structure des cristaux. Paris, 1784.

M. L. FRANKENHEIM: System der Krystalle. 1842.

CHR. WIENER: Grundzüge der Weltordnung. 1863.

A. BRAVAIS: Études cristallographiques. Paris, 1866.

A. KNOP: Molecularconstitution und Wachsthum der Krystalle. Leipzig, 1867.

L. SOHNCKE: Entwickelung einer Theorie der Krystallstructur. Leipzig, 1879.

P. GROTH: Die Molecularbeschaffenheit der Krystalle. Munich, 1888.

O. LEHMANN: Die Molecularphysik. 2 vols. Leipzig, 1888.

A. FOCK: Einleitung in die chemische Krystallographie. Leipzig, 1888.

L. WULFF: Ueber die regelmässigen Punktsysteme—Zeitschrift für Krystallographie und Mineralogie, vol. XIII. pp. 503-566. 1887.

L. SOHNCKE: Erweiterung der Theorie der Krystallstructur—*ibid*. vol. XV. pp. 426–446. 1889.

II. On Crystallography.

R. J. HAUY: Traité de cristallographie. 2 vols. 1822.

F. E. NEUMANN: Beiträge zur Krystallonomie. 1823.

C. F. NAUMANN: Lehrbuch der reinen und angewandten Krystallographie. 2 vols. 1830.

W. H. Miller: A Treatise on Crystallography. 1839.
C. F. Naumann: Elemente der theoretischen Krystallographie. 1856.
V. von Lang: Lehrbuch der Krystallographie. 1866.
A. Schrauf: Lehrbuch der physikalischen Mineralogie. Vol. I. 1866.
F. A. Quendstedt: Grundriss der bestimmenden und rechnenden Krystallographie. 1873.
Rose and Sadebeck: Elemente der Krystallographie. 1873.
C. Klein: Einleitung in die Krystallberechnung. 1876.
E. Mallard: Traité de cristallographie géometrique et physique. Vol. I. 1879.
Th. Liebisch: Geometrische Krystallographie. 1881.
H. Bauerman: Systematic Mineralogy. 1881.
E. S. Dana: Text-book of Mineralogy, 2d Ed. 1883.
A. Brezina: Methodik der Krystallbestimmung. 1884.
P. Groth: Physikalische Krystallographie, 2d Ed. Leipzig, 1885.
V. Goldschmidt: Index der Krystallformen der Mineralien. Vol. I. Berlin, 1886.
F. Henrich: Lehrbuch der Krystallberechnung. Stuttgart, 1886.
V. Goldschmidt: Ueber Projection und graphische Krystallberechnung. Berlin, 1887.
M. Websky: Anwendung der Linearprojection zum Berechnen der Krystalle. Berlin, 1887.
G. Wyrouboff: Manuel pratique de cristallographie. Paris, 1889.

III. On Crystal Aggregates and Irregularities.

A. Weisbach: Ueber die Monstrositäten tesseral krystalisirender Mineralien. 1858.
C. Klein: Ueber Zwillingsverbindungen und Verzerrungen. Heidelberg, 1869.
A. Sadebeck: Angewandte Krystallographie. Berlin, 1876.
G. Tschermak: Zur Theorie der Zwillingskrystalle—Mineralogische und petrographische Mittheilungen, vol. II. p. 499. 1879.
E. Mallard: Sur la théorie des macles—Bulletin de la société minéralogique de France, vol. VIII. p. 452. 1885.
H. Baumhauer: Das Reich der Krystalle. Leipzig, 1889.

CRYSTALLOGRAPHY.

CHAPTER I.

CRYSTAL STRUCTURE.

The Crystal. All chemically homogeneous substances, when they solidify from a state of vapor, fusion, or solution, tend to assume certain regular polyhedral forms. This tendency is much stronger in some substances than in others, and it varies widely in the same substance under different physical conditions.

The regularly bounded forms thus assumed by solidifying substances are called *crystals*.* Their shapes are directly dependent on the nature of the substance to which they belong, and they are therefore valuable for its identification, like any of its physical properties.

Crystal forms have been particularly useful as a means of recognizing and classifying the mineral

---

* This term, which we now apply to all of these forms, was used by the ancients exclusively for crystallized silica or quartz, in allusion to the then accepted idea that this substance was ice rendered permanently solid by the action of intense cold. (Greek, κρύσταλλος, from κρύος, frost; Latin, *crystallus*.) This theory was still accepted by Paracelsus, and was not combated until the beginning of the seventeenth century.

substances which compose the earth's crust; and hence an accurate knowledge of their geometrical and physical properties has long been considered as indispensable to the mineralogist, geologist, and mining engineer. Now, however, such a knowledge is hardly less useful or necessary to the chemist or physicist, irrespective of any interest he may have in the minerals and rocks.

The regular external form of a crystal is its most striking feature, and the only one that, for a long time, was regarded as important or essential. But we now know that this form is only an outward expression of a regular internal structure. A study of the physical properties of a crystal aids us very much in properly understanding the meaning of its external form.

If we examine ordinary homogeneous substances which are not crystals in regard to their physical properties, such for instance as their elasticity, hardness, cohesion, light-transmission, heat-conduction, etc., we find that these are equal in all directions. Thus a piece of glass, when struck, will break with equal readiness along all surfaces, and it will exhibit an equal degree of hardness wherever it may be scratched.

With crystals, however, this is not true. In them we find *differences* of elasticity, hardness, cohesion, and other physical properties, which do *not* exist in homogeneous substances which are not crystals. As the result of long study by many eminent observers, the fact has been established that the distribution of physical properties, like those above enumerated, *is, in crystals, equal along all parallel directions, while, with cer-*

tain exceptions, it is unequal along directions which are not parallel. This important fact gives us the clue to the essential nature of the crystal; for it implies that both the regular external form and the distribution of physical properties are alike directly the outcome of some regular internal structure. We may make a glass model of exactly the same shape as a crystal, but it is not a crystal, in spite of its form, because the necessary internal structure is absent.

Crystallography. In its broadest sense this term relates to the scientific description of crystals in all their aspects. This wider usage, however, naturally falls into three subdivisions—*geometrical* or *morphological*, *physical*, and *chemical* crystallography. Of these three subjects only the first, for which the general term crystallography is still in a stricter sense reserved, is embraced within the scope of this book.

The regular forms exhibited by crystals were made a subject of elaborate study, before the distribution of their physical properties received attention. In this way these forms were found to obey certain laws which rendered their mathematical treatment and classification possible. Hence the science of geometrical crystallography developed quite independently.

Now, however, that physical crystallography has shown the complete accord between the forms and physical behavior of crystals, as well as the direct dependence of both on a regular internal structure, we must, if we would fully appreciate the real significance of crystal form, first discover what we can of the nature and mode of arrangement of the crystal particles, themselves invisible.

The Elementary Crystal Particles. The modern conception of matter is that it is composed of ultimate particles, called *atoms*, which are always in a state of intense vibration, and which are separated from each other by distances vastly greater than their own diameters. The number of kinds of these atoms is comparatively small, but by uniting into groups of varying size, composition, and arrangement (called *chemical molecules*), they are capable of producing all the variety of substances which compose our material world.

It is not improbable that, as the ultimate elements of matter (the atoms) unite to form chemical molecules, so these, in their turn, may unite to form groups of a higher order, called *physical molecules*. The former may be thought of as comparable to our solar system, composed of individual planets which are united to a single group by the attraction of gravitation, while at the same time the solar system as a whole is but a single unit in a vastly larger group of systems, which is held together by the same attractive force.

The chemical molecule is the *unit of substance*, because we cannot imagine it to be divided without altering the substance. The physical behavior of crystals also necessitates *units of structure*, or elementary particles by whose regular arrangement the crystal is built up. As such units we may assume the physical molecules; and, for this purpose, it is immaterial whether they are different from the chemical molecules or not.

Whatever the true size and nature of these crystal units (crystal molecules, physical molecules *) is, we

---

* The word *molecule*, as used in the following pages of this book, should be understood as always referring to the physical and not to the chemical molecules.

can, for all purposes of explaining crystal structure, regard them as points (their centres of gravity) surrounded by ellipsoids or spheres whose size and form represent the sum of all the various attractive and repellent forces inherent in the molecules. All the crystal molecules of the same chemical substance under the same conditions must be identical in size, shape, and in the distribution of forces; for different substances they must be different, while for the same substance under different conditions they may or may not be different.

Mode of Molecular Arrangement in Crystals. If the crystal elements or physical molecules of a given substance possess the same size and the same attractive forces, then, in case these molecules are perfectly free to act and react upon each other, they must all assume a similar position relative to one another, i.e., such a position that equivalent directions of attraction and repulsion in all the molecules shall be parallel.

To illustrate this, let us assume, as the simplest possible case, that the distribution of attractive forces in the physical molecules of a certain substance is equal in three directions at right angles. Then such a molecule may be graphically represented by Fig. 1, where the three equal and perpendicular dotted lines represent the intensity and direction of the attractions inherent in the molecule. Now if a great number of similarly constituted molecules of this kind are gradually approaching, while their forces are entirely free to react upon each other, they will finally arrange themselves in parallel positions, and at equal distances, corresponding to the cor-

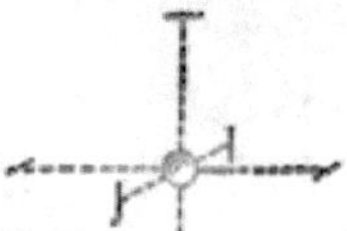

Fig. 1.

ners of a cube, Fig. 2. If we conceive of an arrangement like this as indefinitely extended, we see that *the grouping about any molecule must be the same as about every other;* and also that *the arrangement of molecules in all parallel planes, and along all parallel lines, must be the same.* This is still more distinctly seen in Fig. 3, where the distribution of molecules along the line $e_1 c_2 a_3$, is evidently different from that along $c_1 b_2 a_3$, or $a_1 a_2 a_3$; so also the arrangement of molecules in the plane $e_1 a_3 e_9$ is different from that in the plane $a_1 a_7 c_3 c_9$, etc.; while in parallel planes, like $a_1 a_2 a_3 a_4, b_1 b_2 b_3 b_4$, etc., the grouping is the same.

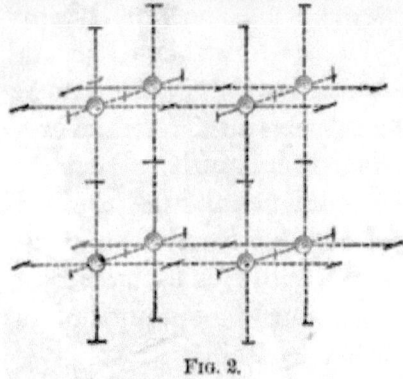

Fig. 2.

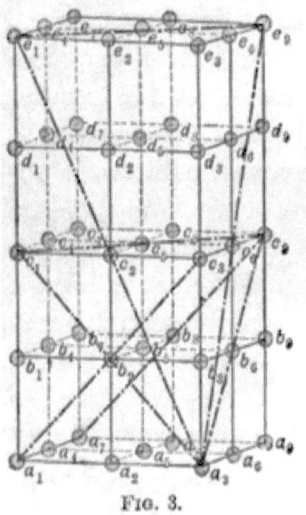

Fig. 3.

No arrangement of molecules, in which they are *not* similarly distributed in all parallel directions, or where the grouping is *not* the same about each, can be regarded as possible in a crystal. Thus the study of crystal structure becomes an investigation of the possible networks of points in space which satisfy these conditions. This problem has been dealt with by various writers, especially by Sohncke, who finds that all such arrangements which are possible (sixty-six in number) fall naturally into groups whose symmetry corresponds with that of the systems to which all crystal forms belong.*

Crystalline and Amorphous Substances. If the passage of a chemically homogeneous substance from the gaseous or liquid into the solid state is too rapid to allow of the perfect action of the attractive and repellent forces upon each other, the crystal molecules may become fixed while their parallel orientation is still incomplete or even while it is wholly wanting.

Many substances show by their physical properties that they possess no regularity of molecular structure whatever. Such substances never exhibit characteristic polyhedral forms, and are therefore said to be *amorphous*. Substances which are only known in the amorphous state are usually of indefinite chemical composition, like coal, amber, or opal. Definite chemical compounds almost always possess *some* power to crystallize, though certain usually crystallized substances may be made to assume an amorphous form by very much accelerating their rate of solidification, e.g., many silicates, when fused and rapidly cooled, form a glass. The real difference between amorphous and crystalline substances is therefore internal and molecular. A fragment of quartz and a fragment of glass may to all external appearances be quite alike, but

---

* Those desiring further information on this subject should consult the works of Frankenheim, Bravais, Sohncke, Groth, and Wulff, cited on p. vi.

Sohncke has recently been led to extend his original theory by the assumption that, in certain cases, there may be in a single crystal two or more interpenetrating networks like those above described. (Zeitschrift für Krystallographie, vol. xiv. p. 426; 1888.)

they still possess important internal differences. In the former the elasticity is equal in parallel directions and different in directions not parallel, while in the latter the elasticity is equal in all directions.* In the former the molecular arrangement is regular; in the latter quite irregular.

If we symbolize the physical molecule by a sphere with all of its attractive and repellent forces resolved into three directions not at right angles to one another, the position of such molecules in different states of matter may be graphically represented by the following four diagrams. Fig. 4 shows the spheres so widely separated as to be wholly without each other's influence and therefore free to move in any direction, as

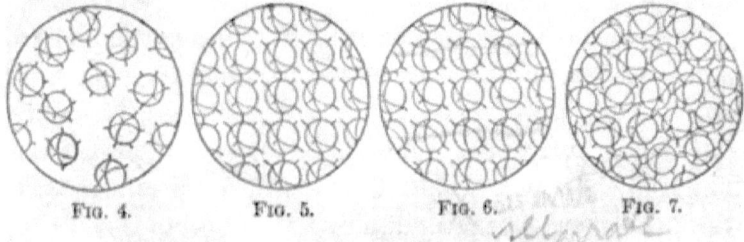

Fig. 4.   Fig. 5.   Fig. 6.   Fig. 7.

in a gas or liquid. Fig. 5 represents a solid state, where each molecule is free to move only within its own sphere of attraction, the position of this being conditioned by those of the surrounding molecules. In this case the orientation of the molecules is complete, as in a simple crystal. Fig. 6 illustrates another solid state where the orientation is only partial, as in the case of twin crystals. Here the molecules of each

---

* This property of amorphous bodies does not, of course, apply to organized substances, like wood, which are not strictly homogeneous, nor to others whose differences in elasticity are due to external strain, as in the case of unannealed glass.

horizontal row have two of their axes parallel with those above and below them, but not the third. Such a position would be reached by supposing the molecules of alternate rows in Fig. 5 to have been revolved 180° about a normal to the plane of the paper. Fig. 7 represents a solid state where there is no regularity of molecular orientation, as in the case of amorphous substances.

Strength of the "Crystallizing Force." The molecular forces which tend to produce a regular internal structure in matter as it slowly solidifies exert themselves in varying degrees, both in different substances under the same conditions and in the same substance under different conditions. Such variations may be called differences in the strength of the "crystallizing force."

This crystallizing force, while it probably exists to some degree in all substances of definite chemical composition, is so very weak in certain ones, like serpentine or turquoise, that their crystal form is not definitely known. That they really crystallize, and are not, strictly speaking, amorphous bodies, is shown by their optical and other physical properties, although conditions favorable enough for the production of their crystal form appear never to be fulfilled.

Other substances, like some of the metallic sulphides, rarely possess a well-defined crystal form. They commonly occur in what is termed the *massive state*, i.e., in crystalline aggregates which show little or no trace of crystal planes.

Still other substances, like calcium carbonate (calcite), and silica (quartz), possess an intensely strong crystallizing force, and are rarely found except in well-defined crystals.

Mode of Crystal Growth. Crystals are distinguished from living organisms by the method of their growth. While the latter grow from within outward and are conditioned both in their form, size, and period of existence by the internal laws of their being, crystals enlarge by regular accretions from without, and are limited in size and duration only by external circumstances.

Organisms must pass through a fixed cycle of constantly succeeding changes. Youth, maturity, and old age are unlike and must come to all in the same order. There is, furthermore, in nearly all living things a differentiation of organs, limitation in the extent of growth, and the power of reproduction.

In crystals, on the other hand, every part is exactly like every other part. Our very definition of crystal structure is an arrangement of particles, the same about one point as about every other point; hence, in one sense, the smallest fragment of a crystal is complete in itself.

Moreover, since crystals grow by the addition of regular layers of molecules, arranged just like all other layers, we can set no limit to the size of a crystal, so long as the supply of material and conditions favorable to its formation remain constant. There is in fact the widest divergence in the size of crystal individuals of the same composition and structure. Those of ultra-microscopic dimensions and those many feet in length may be identical in everything but size. Both are equally complete, and one is in no sense the embryo of the other. As a rule, the size of a crystal is inversely proportional to its purity and perfection of form, but this, as will be seen at once, is dependent on external conditions.

Finally, the individual crystal, unlike the individual organism, will remain unchanged so long as its surroundings are favorable to its existence.

Crystal Habit. Since the growth of a crystal is produced by the addition of regular layers of molecules, then, if at any time this growth be interrupted, the crystal will be bounded by plane surfaces which represent the position of such molecular layers. The particular planes possible in any given case must therefore depend upon the mode of molecular arrangement, and hence upon the chemical composition of the crys-

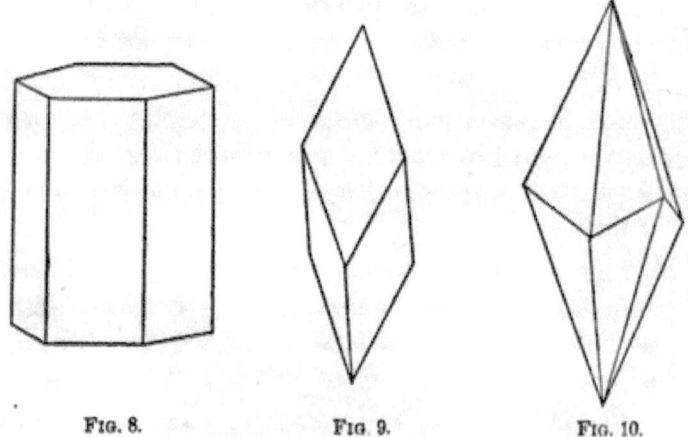

FIG. 8.   FIG. 9.   FIG. 10.

tal. Nevertheless the number of planes *possible* on a crystal is very much greater than that which actually occurs in any single instance. It is reasonable to suppose that only planes passed through a regular network of molecules so as to intersect the same number at equal distances along all parallel lines are possible crystallographic planes; while those planes will be of the most frequent occurrence which intersect, in this way, the greatest number of molecules. Thus we may have crystals identical in composition and in all their physical properties, but bounded by very different sets of planes, all of which are equally possible with the same internal structure. Such differences in form among crystals of the same substance condition what is known as *crystal habit*.

The three preceding figures, 8, 9, and 10, represent crystals of calcium carbonate (calcite). The forms, though apparently so unlike, can all be shown to be derived from the same molecular arrangement. Each presents a combination of certain out of a great number of possible planes, and therefore exhibits a particular *habit* of a single substance.*

---

* A graphic illustration of the molecular structure, as well as of the habit of crystals, may be advantageously employed, which, in principle, is not unlike the well-known figures of Haüy. If we represent the physical molecules by any small spherical bodies, of nearly the same size, like shot, we can readily see how it is possible, by the same arrangement of these, to build up different forms. A square of such bodies, arranged in parallel rows, may be taken as a starting point, and then by piling others upon them, as cannon-balls are piled, a symmetrical four-sided pyramid is produced. If the shot be made to cohere by dipping them in shellac, a similar pyramid may be built up on the other side of the base, thus forming the regular octahedron. Again, if successive horizontal and vertical layers be taken away equally from each of the six solid angles of the octahedron, this form is seen to develop gradually into the cube, while the interior structure remains unchanged. Finally, we may use each of the six faces of the cube as a base for the erection of a quadratic pyramid, and thus the dodecahedron is formed, with a structure like that which produced the other two figures.

Such models as these admirably illustrate how differences of habit may result from the same molecular arrangement, as well as how the planes of one form may replace the edges or angles of another. We have only to conceive of the shot as too small to be visible, and the surface produced by any layer becomes a crystal plane.

Exactly what it is that determines the habit of a crystal is not known. Crystals formed at the same time generally exhibit the same habit, but this is not always the case. Doubtless many slight alterations in external conditions at the time of formation may be influential in this regard. Certain crystals possess different habits at different periods of their growth. In the case of transparent substances a core or kernel, bounded by different planes from those on the exterior of the crystal, is sometimes visible in the interior (German, *Kernkrystall*). Fig. 11 represents a calcite crystal of rhombohedral habit, from Mineral Point, Wisconsin, in the centre of which is a darker scalenohedron. Fluorspar also shows this phenomenon. Experiments with many artificial salts appear to indicate that the presence of impurities in the concentrated solution may be a most important factor in conditioning the habit of crystals. For example, sodium chloride may be obtained in octahedrons, instead of the usual cubes, when crystallized from a solution containing sodium hydroxide. On the other hand, alum, which usually crystallizes in octahedrons, can be produced in the form of cubes from alkaline solutions. The habit of crystals of Epsom salts (magnesium sulphate) is also modified by the presence of borax in the solution. (See O. Lehmann, Molecularphysik, vol. I. p. 300.)

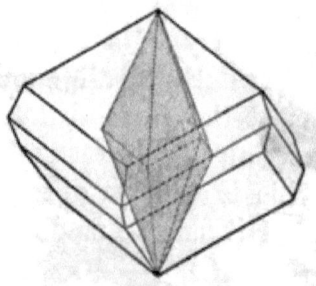

Fig. 11.

Another most important variation in crystal habit is often produced by what is known as *distortion* of the form. While crystals are increasing in size by the addition of layers of molecules, it will rarely happen that the concentration of the mother-liquor will be so evenly balanced on all sides as to make the growth equally rapid in all directions. Where the most material is supplied, there the growth will be most accelerated; while the size of a given plane will be relatively diminished in proportion as it grows from the centre of the crystal. It thus happens that crystallographically equivalent planes vary much in size on the same individual, and that, in this way, the symmetry of a form is often completely disguised; to restore it, all similar planes must be imagined as at the same distance from the centre. In this way an *ideal form* is derived from the distorted form. This is a matter of so much importance to the beginner that it may be made the subject of illustration.

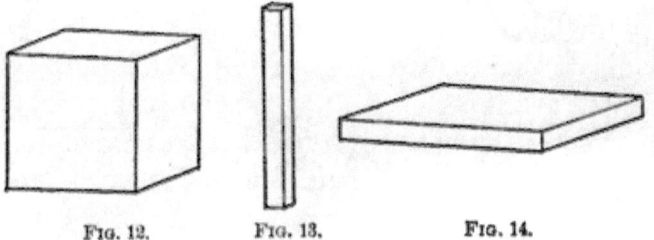

Fig. 12. Fig. 13. Fig. 14.

The cubic crystal, Fig. 12, may grow most rapidly in one direction, becoming prismatic, Fig. 13; or in two directions, becoming tabular, Fig. 14, without losing its character as a cube so long as its angles remain 90°.

The symmetrical octahedron, Fig. 15, may become distorted, as in Fig. 16, or even flattened into triangular plates, Fig. 17, as is frequently the case with alum crystals.

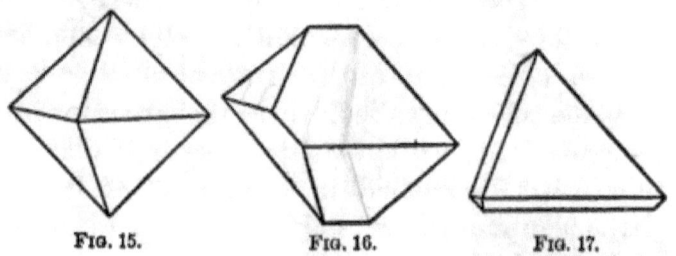

FIG. 15.   FIG. 16.   FIG. 17.

The next three figures, 18, 19, and 20, show the same combination of planes unequally developed on three crystals of quartz.

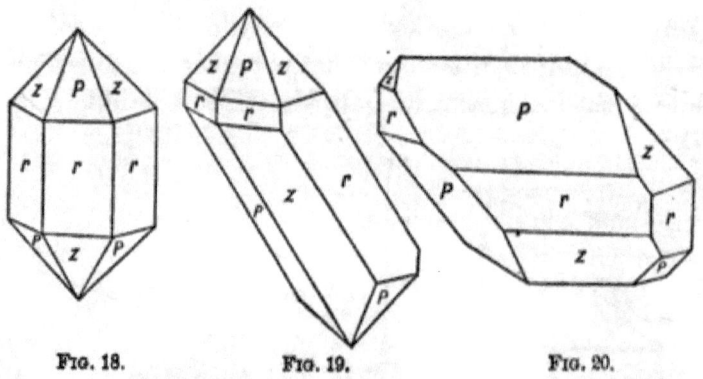

FIG. 18.   FIG. 19.   FIG. 20.

The occurrence of ideal forms in nature is rather the exception than the rule; hence the constructing in imagination of the symmetrical equivalents of more or less distorted crystals becomes an important matter for practice. For this purpose it is quite necessary that the student first familiarize himself with the ideal forms; and crystal models, with which the study of crystallography must be commenced, are on this account generally represented as symmetrical bodies.*

Crystal Individuals and Crystalline Aggregates. That portion of a homogeneous crystallized substance whose molecular arrangement is throughout the same along all parallel lines, and which is bounded by its own characteristic plane surface, is called a *crystal individual*. Such an individual is not of necessity completely bounded by crystal planes, since there is generally a larger or smaller point of attachment to other crystals. There must, however, be enough planes to allow of the restoration of the complete form. Anything less than this is a crystal fragment or grain.

The union of two or more crystal individuals produces a *crystal aggregate;* while a mass of crystal grains, devoid of their characteristic forms and closely packed together, may be termed a *crystalline aggregate*.

---

* Models, which are of such prime importance in the study of crystallography, are most cheaply and elegantly made in Germany. The principal forms in all systems are constructed of glass by F. Thomas at Siegen in Westphalia. These show the position of the axes inside by colored threads, and are particularly valuable for demonstrating the derivation of hemihedral and tetartohedral forms as well as different methods of twinning. They are large enough to be well suited for class-instruction ; and, considering the perfection with which they are made, are furnished at a very reasonable price.

Models of convenient size are accurately made of hard wood by various German firms. The best may be had of Krantz in Bonn, who will furnish collections of any required size. W. Apel of Göttingen also manufactures a small but useful set. Catalogues of all these firms may be had on application. (See also the price-list of crystallographic apparatus at the end of Groth's Physikalische Krystallographie; 2d ed., 1885.)

Cardboard models may be made by students by cutting out and bending into shape the outlines furnished by various authors. (See **Kopp**, Einleitung in die Krystallographie; 1862. Atlas.)

CRYSTAL STRUCTURE. 17

This distinction may be made clearer by the two following figures, 21 and 22.

FIG. 21.—CRYSTAL AGGREGATE.
(Quartz from Dauphiné, France.)

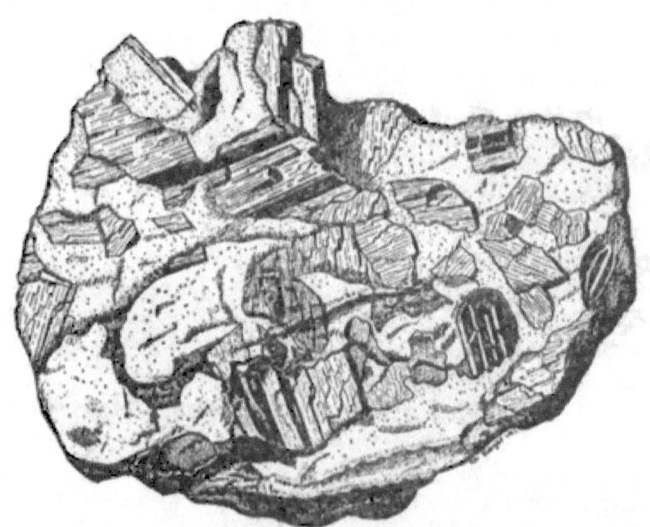

FIG. 22.—CRYSTALLINE AGGREGATE.
(Gabbro from Prato near Florence, Italy.)

Aggregates may further be *homogeneous* and *heterogeneous*, according as they are composed of one substance with but one kind of molecular structure, like marble, or of two or more substances with different internal structures, like granite. The subject of crystal aggregates is more fully treated of in Chap. IX.

Limiting Elements of Crystals. With certain exceptions to be explained beyond, all crystals have their limiting planes so arranged in *pairs*, that to every one there is a parallel plane on the opposite side of the crystal.

The planes, or *crystal faces*, intersect in *edges* and *angles*. A crystal edge is the line of intersection of two crystal planes; the angle which such an edge encloses is called an *interfacial angle*. By the term *crystal angle* is meant the solid angle in which three or more crystal faces meet.

Similar edges are those in which similar planes intersect at equal angles; *similar angles* are those enclosed by the same number of planes, similarly arranged and meeting at the same inclination.

It is convenient to remember that on all polyhedrons, and hence on all crystals, the number of faces plus the number of solid angles is equal to the number of edges plus two.

$$F + A = E + 2.$$

General Principles of Crystallography. Before proceeding to the description of the different groups or systems into which all crystal forms are classified, it will be necessary to consider certain common properties which such forms possess. These are:

1. Constancy of corresponding interfacial angles on all crystals of the same substance.
2. Simple mathematical ratio existing between the

co-ordinates of all planes which are possible on crystals of the same substance.

3. Symmetry.

We may regard the expression of these common characters of crystals as the *fundamental laws* of crystallography, and their explanation will form the subject of the following chapter.

CHAPTER II.

GENERAL PRINCIPLES OF CRYSTALLOGRAPHY.

1. LAW OF THE CONSTANCY OF INTERFACIAL ANGLES.

Statement of the Law. *However much the crystals of the same substance may vary in habit and in the relative size and development of similar planes, their corresponding interfacial angles remain constant in value; provided that, first, they possess identically the same chemical composition, and, second, that they are compared at the same temperature.*

This identity of corresponding angles on crystals of the same substance is clearly a necessary consequence of their possessing the same molecular structure under the same physical conditions. It was, however, observed to be true long before crystals were thought to have any peculiar structure, and was first formulated by Steno in 1669. It was further substantiated by Romé de l'Isle with a contact-goniometer in 1783, and finally established within a very small limit of error, after the invention of the reflecting goniometer in 1809.

The importance of this law consists in showing that the exact values of crystal angles, even more than the particular shapes of the crystals themselves, are characteristic of the substance composing them, for these angles remain constant in spite of all the distortions to which the forms are subject (p. 14).

Indeed, they may serve to distinguish substances which are in all other outward respects quite identical. For instance, the angle 105° 5' is always included between the cleavage faces of a crystal of calcium carbonate, while the corresponding angle on crystals of magnesium carbonate is 107° 28'.

The essential nature of the interfacial angles renders more intelligible the distortion of crystal forms described in the preceding chapter, since, in spite of a parallel shifting of the planes, the angles between them remain unaltered.

Measurement of Interfacial Angles. In order to ascertain how far the law of the constancy of interfacial

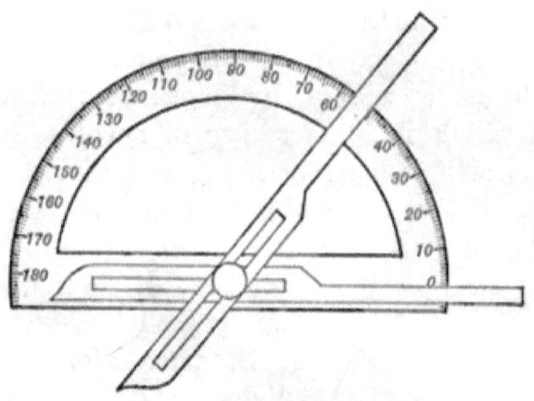

FIG. 23.

angles holds good it is necessary to accurately measure these angles. A knowledge of their exact values is furthermore requisite for any mathematical treatment of crystal forms. An instrument for measuring crystal angles is therefore of prime importance in the study of crystallography, and is called a *goniometer*.

The simplest form of such an instrument, called a *contact-* or *hand-goniometer*, was first constructed near the end of the last century. It consists of two arms (Fig. 23), one of which revolves about a pivot fastened to the other, which may be set by a screw at any angle. These two arms are applied to the two faces of the crystal whose interfacial angle is to be measured, and, when as nearly in contact as possible, the screw is set and the angle read by placing the arms upon a graduated arc. Such measurements are not reliable to within less than half a degree. They are chiefly valuable for large crystals whose faces are rough or unpolished.

For accurate measurements of the interfacial angles of crystals we must have recourse to the *reflection-goniometer*, whose construction depends upon a principle first made use of by Wollaston. The angle through which it is necessary to revolve a crystal about one of its edges, so as to successively obtain a reflection of the same object from the two planes whose intersection forms the edge, is equal to the supplement of their interfacial angle.

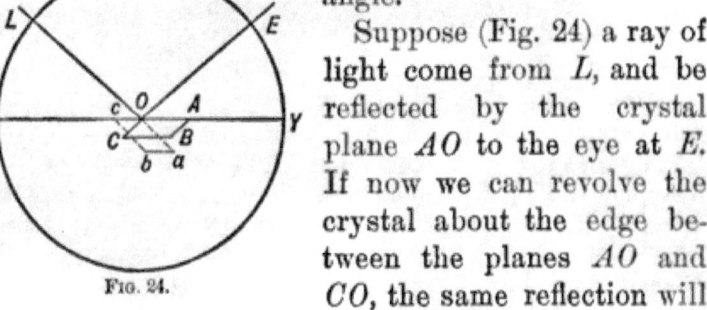

Fig. 24.

Suppose (Fig. 24) a ray of light come from L, and be reflected by the crystal plane AO to the eye at E. If now we can revolve the crystal about the edge between the planes AO and CO, the same reflection will be sent to the eye, stationary at E, from the plane CO, when it has been brought exactly into the position of AO. In order to secure this, the crystal must obviously be revolved through the angle COc, which is the supplement of AOC, the interfacial angle required.

For the successful application of this principle to practical measurement four things are necessary:

1. The eye must be kept at the same point.

2. The object reflected must be at a sufficient distance to make the rays coming from it practically parallel.

3. The edge to be measured must be parallel to the axis of revolution, and normal to the plane of the graduated circle upon which the angle is read. (*Adjustment.*)

4. This edge must also lie exactly in the continuation of the axis of revolution. (*Centering.*)*

2. LAW OF THE SIMPLE MATHEMATICAL RATIO.

Crystallographic Axes. In order that we may be able to classify and compare the forms of crystals, we must have some ready and simple mode of expressing the

---

* The description of the various forms of reflection-goniometers, together with all the detail of their construction and successful manipulation, lies far beyond the scope of this book. For information on this subject, the student may consult:

E. S. Dana, Text-book of Mineralogy; 2d ed., 1883; pp. 83–88 and 115–118.

P. Groth, Physikalische Krystallographie; 2d ed., 1885; pp. 560–584.

A good reflection-goniometer is quite indispensable to all who intend to do any special work in crystallography. The best, with horizontal circles, are manufactured by R. Fuess, 108 Alte Jakobstrasse, Berlin, at prices ranging from $80 to $350, according to the completeness of their equipment. The most serviceable for all ordinary purposes is his Model II, costing about $150. A much simpler instrument with vertical circle, constructed on the old Wollaston plan, is made by Voigt & Hochgesang of Göttingen.

relative positions and inclinations of their planes. This is accomplished by referring them to systems of axes, according to the method of analytical geometry.

The position of any crystal plane is thus fixed by and expressed in the relative lengths of its intercepts on the axes to which it is referred. The axes to which the planes of a crystal are referred, called the *crystallographic axes*, may be of equal or unequal length, and may intersect at either oblique or right angles.

When the crystallographic axes are of unequal length, it is customary to designate the one which stands vertically by the letter c, the one which runs from right to left by b, and the one which runs from front to back by a. The two extremities of each axis are distinguished by the plus or minus sign, as shown in Fig. 25. If all three axes are of equal length, they are all represented by a. If two are of equal length, they are designated by a, and the third one by c.

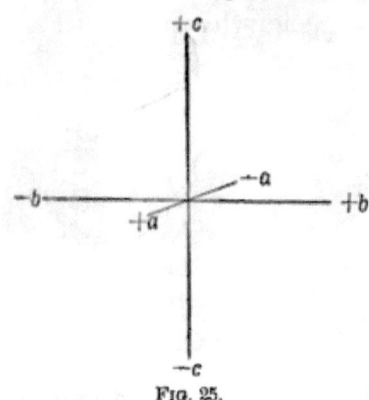

Fig. 25.

If the axial intersections are not rectangular, they are designated by the Greek letters α, β, and γ, as follows: $b \wedge c = \alpha$, $a \wedge c = \beta$, and $a \wedge b = \gamma$. (Fig. 26.)

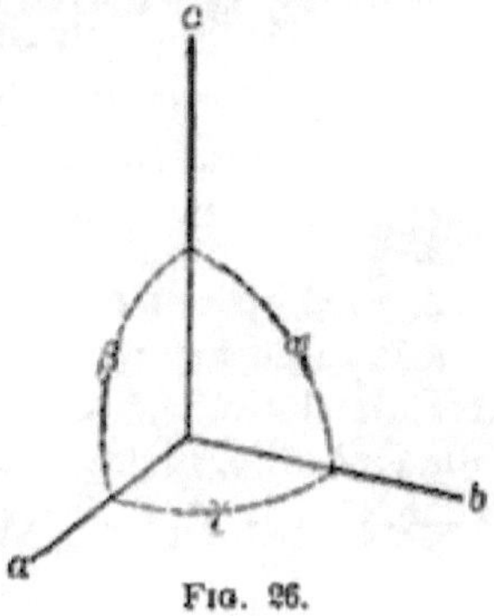

The planes in which two of the crystallographic axes lie are called *axial* or *diametral*

Fig. 26.

planes. They are the coördinate planes of analytical geometry, and divide the space within the crystal into eight sextants, called *octants;* or, in one system where four axes are used, into twelve sextants, called *dodecants.* The axial planes are also sometimes called principal sections (German, *Hauptschnitte*).

Parameters. The values of the intercepts of any crystal plane on the axes are called the *parameters* of the plane. They are expressed in terms of certain axial lengths which are assumed as unity. Suppose (Fig. 27) that ABC is a plane which intersects the axes X, Y, Z at their unit lengths. The position of any other plane, HKL, is determined, if we know the values OH, OK, and OL in terms of OA, OB, and OC. In this case $\dfrac{OH}{OA} = 2$;

Fig. 27.

$\dfrac{OK}{OB} = 1$; $\dfrac{OL}{OC} = \dfrac{1}{2}$. These quotients are the *parameters* of the plane HKL.

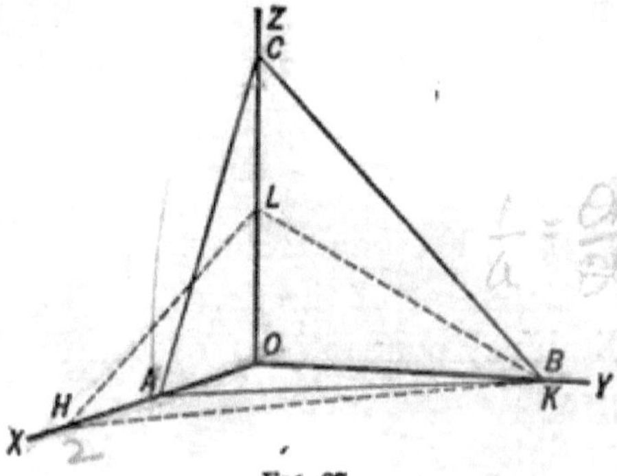

If we denote the axes X, Y, Z by a, b, and c, the most general symbol for any plane becomes

$$na : pb : mc,$$

where n, p, and m are rational quantities and the parameters of the plane.

Since, however, any plane may be thought of as shifted in either direction, so long as the *relative* value of its intercepts remains the same (p. 14), one of the three parameters may always be made equal to unity, and the most general expression for any crystal plane becomes

$$na : b : mc.$$

Indices. The position of a crystal plane may be equally well expressed by employing the reciprocals of the parameters, which are called the *indices*. For purposes of notation these possess many practical advantages over the parameters, and are therefore quite generally used. We can readily see from Fig. 27 that the plane HKL may be located by the values $h = \dfrac{OA}{OH}$; $k = \dfrac{OB}{OK}$; $l = \dfrac{OC}{OL}$, where h, k, l are the reciprocals of n, p, m, the parameters.

Statement of the Law. With the aid of these preliminary conceptions of what is meant by crystallographic axes, parameters, and indices, we may formulate the law of the simple mathematical ratio (also known as the law of the rationality of the indices) as follows: *Experience shows that only those planes occur on any crystal, whose axial intercepts are either infinite or small, even multiples of unity.* The ratio of intercepts on the same axis for all planes possible on the same crystal is therefore rational, and may be expressed by small whole numbers, simple fractions, infinity or zero.*

---

* In certain cases disturbances during the growth of a crystal, or other causes not understood, produce surfaces whose intercepts are very large, if not almost irrational. Websky has called such faces *vicinal planes*. Their significance is not entirely clear, and their consideration lies outside of the scope of this work.

GENERAL PRINCIPLES. 27

This law leads us to the same result as was reached by the preceding law of constant interfacial angles, viz., that every plane cannot occur on a given crystal. It brings us, however, the additional information that those planes which can occur must have rational axial intercepts. This law, like its predecessor, is a necessary deduction from a regular molecular structure, since it can be mathematically demonstrated that only those planes which possess rational indices satisfy the conditions of a possible crystal plane (p. 11.) Thus observation corroborates our hypothesis of molecular structure.

Systems of Crystallographic Notation. In order to compare crystal planes, we must be able to designate them by generally applicable systems of symbols. Many such systems have been devised,* but those now most generally employed aim to locate the position of each crystal plane with reference to the crystallographic axes, and are therefore based upon the use of either parameters or indices.

Parameter System of Weiss. This is one of the oldest as well as one of the most easily understood of all systems of crystallographic notation. The three axes are written in the fixed order explained on page 24: $a:b:c$ if of unequal length; $a:a:c$ if two (then made the lateral axes) are of equal length; or, $a:a:a$ if all three are of equal length. To each axial letter is then prefixed the numerical value of its parameter,

---

* Those desiring a full description and comparison of all the different systems of crystallographic notation which have been suggested, will find it in Goldschmidt's Index der Krystallformen, vol. I.

whenever this is not unity. In the most general symbol for any plane, $na : pb : mc$, it is customary to reduce the value of one of the two lateral axes, a or b, to unity (p. 26); the parameter of c does not, however, become unity unless the parameter of one of the lateral axes is at the same time unity. For this reason, the most general symbol for any plane in the notation of Weiss is $na : b : mc$ or $a : nb : mc$; but not $na : mb : c$. Hence the value of the parameter n can vary only between one and infinity, because whichever of the lateral axes (a or b) is the shorter is assumed as unity; m, on the other hand, varies between zero and infinity, because it refers to the single axis c. For example, a plane whose axial intercepts were $\frac{1}{2}$ on a, $\frac{1}{3}$ on b, and 1 on c, would not be written $\frac{1}{2}a : \frac{1}{3}b : c$, but $\frac{3}{2}a : b : 3c$; so again, a plane whose intercepts were 1 on a, $\frac{1}{2}$ on b, and $\frac{1}{3}$ on c would be written, $2a : b : \frac{2}{3}c$, etc. If all three axes are of equal length, and therefore interchangeable, none of the parameters ever becomes less than unity.

Parallelism of any plane to a crystallographic axis is indicated by Weiss with the sign of infinity, ∞, written in its proper place as a parameter. Thus the symbol of a plane parallel to the a axis, but cutting b and c at unity, becomes $\infty a : b : c$; the symbol of a plane parallel to c and one of the lateral axes, becomes $\infty a : b : \infty c$ or $a : \infty b : \infty c$. Since, however, one of the two lateral axes must always be unity, the symbol for a plane parallel to both lateral axes is not written $\infty a : \infty b : c$, but $a : b : 0c$.

This notation is employed in the writings of Weiss, Rose, Quendstedt, Rammelsberg, and some other authors; but, on account of its cumbersome character, it has generally given place to shortened forms, suggested by Naumann and Dana.

Abbreviated Parameter Symbols. Naumann has proposed a system of symbols, now in quite general use, which contain in their centre certain capital initials —O (octahedron) when the planes are referred to a set of equal axes, and P (pyramid), when they are referred to systems of unequal axes.

Before this letter is written the parameter m, which refers to the vertical axis, and *after* it the parameter n, which belongs to one of the lateral axes. Any parameter whose value is unity is omitted. Thus the most general symbol becomes mPn. P or O signifies a form whose planes cut all three axes at unity; mP, a form whose lateral axes are both unity; and Pn, a form whose vertical axis is unity and one of whose lateral axes is n. Parallelism to any axis is represented by the sign of infinity, ∞, in its proper place in the symbol. $\infty P \infty$ signifies a plane parallel to the vertical and one lateral axis; but the symbol $P\infty\infty$ ($\infty a : \infty b : c$) is replaced by its equivalent, OP ($a : b : Oc$), for a plane parallel to both lateral axes, as in the notation of Weiss. Many other points regarding the notations of Weiss and Naumann will be brought out in the course of the descriptions of the forms of the various crystal systems.

J. D. Dana has suggested a further simplification of Naumann's symbols, which has come into current use in the United States. It consists in substituting a hyphen for Naumann's initial, and the letter i or I for the sign of infinity. The fundamental form is represented by 1; a single parameter is written alone, if it refers to the vertical axis, c; but it is preceded by 1-

if it refers to one of the lateral axes. Thus Naumann's symbol 2O2 or 2P2 becomes, in Dana's notation, 2-2; 2P, 2; P3, 1-3; P, 1; ∞P, I; P∞, 1-∞; ∞P∞, i-i; OP, O; 3P⅔, 3-⅔; etc. These symbols can present no difficulty to any one familiar with those of Naumann.

Index System of Miller. A system of crystallographic notation based upon the use of the indices owes its present wide application to the writings of Prof. W. H. Miller. Hence it is called the Miller system, although it was first devised in 1825 by Whewell.

In this system the symbol of any plane consists of the reciprocals of the parameters, written in the order above given for the axes a, b, c, and in the simplest form possible when they are cleared of fractions. This can be made most clear by an example. The symbol $3P⅔$ of Naumann becomes ⅔$a : b :$ 3c in the notation of Weiss. The reciprocals of these parameters are ⅔, 1, ⅓, which, when cleared of fractions, become 2 3 1 (read *two, three, one*). This is the symbol for the same plane in the notation of Miller. The indices are always written in the same order, and without any symbol for the axes or crystal system.

A plane parallel to an axis must, of course, contain the index 0 (the reciprocal of infinity) in its symbol. Thus ∞ P ∞ becomes 100 or 010; 0P, 001; ∞ P, 110; P∞, 011 or 101; etc.*

---

* A modification of the use of indices as crystal symbols has recently (1886) been suggested by Goldschmidt. The Miller symbol is shortened from three terms to two by reducing the third to unity. Goldschmidt's indices, p and q, are $= \frac{h}{l}$ and $\frac{k}{l}$. In this case it is of course impossible to avoid fractions (see Index der Krystallformen der Mineralien, vol. i. p. 12).

It is evident that the symbols of Naumann and Dana stand, not for single planes, but for all the planes which have equal intercepts on equivalent axes. Such an assemblage of planes is technically known as a crystal form (see beyond, p. 35). The symbols of Weiss distinguish between those planes of a form which belong to the same octant, by varying the order in which their parameters are written; they do not, however, ordinarily locate a plane in any particular octant, although this may be done by signs.

It is an advantage to be able to designate any particular plane of a crystal form, and this is done in the notation of Miller by the use of signs. Any index referring to a negative end of an axis (see Fig. 25) has a minus sign written over it. This serves to locate a given plane in one particular octant, thus:

$$hkl \quad \bar{h}kl \quad h\bar{k}l \quad hk\bar{l}$$
$$\bar{h}\bar{k}\bar{l} \quad h\bar{k}\bar{l} \quad \bar{h}k\bar{l} \quad \bar{h}\bar{k}l$$

The members of each pair of parallel planes have the same indices and complementary signs; hence to change the signs of any Miller symbol is to change the plane to its parallel and therefore equivalent plane on the opposite side of the crystal.

Miller writes the indices of a single plane either alone, hkl, or inclosed in parenthesis, (hkl). If an entire form is to be represented, the indices are inclosed in brackets, $\{hkl\}$.*

---

* It is quite essential that the student should become equally familiar with the symbols of Weiss, Naumann (or Dana), and Miller. For this purpose practice in transforming the symbols of one sys-

32 CRYSTALLOGRAPHY.

Lévy's System. The oldest system of crystallographic notation was devised by Haüy and subsequently modified by Lévy and Des Cloizeaux. It is still in general use in France, but, on account of its complicated and difficult character, it has no currency in other countries. No explanation of this system need be attempted here. Further information regarding it may be found, if desired, in the works of Des Cloizeaux, Goldschmidt, and Groth.

3. LAW OF SYMMETRY.

Statement of the Law. It has been found that an important property of crystal forms, and one according to which they may be advantageously classified, is their symmetry.

A *plane of symmetry* may be defined as a plane which is capable of dividing a body into two halves which are related to each other in the same way that an

tem into those of the others is recommended until the matter presents no further difficulty. A few examples for such practice are here appended:

Weiss.			Naumann.	Dana.	Miller.
$a:$	$a:$	a	O	1	{111}
$a:$	$a:\infty a$		∞O	I	(011) (110)
$\infty a:\infty a:$		a	$\infty O\infty$	$i\text{-}i$	(001)
$a:$	$a:$	c or $a:b:c$	P	1	{111}
$\{\infty a:\infty a:$		c or $\}$	OP	0	{001}
$\{\infty a:\infty b:$		$c\ \ \ \}$			
$\infty a:$	$b:$	∞c	$\infty P\infty$	$i\text{-}i$	{010}
$a:$	$\infty b:$	∞c	$\infty P\infty$	$i\text{-}i$	{100}
$\infty a:$	$b:$	c	$P\infty$	$1\text{-}i$	{110} {DM}
$2a:$	$b:$	$2c$	$2P2$	$2\text{-}2$	{121}
$3a:$	$b:$	c	$P3$	$1\text{-}3$	{133}
$a:$	$b:$	$3c$	$3P$	3	{331}
$a:$	$2b:$	$3c$	$3P2$	$3\text{-}2$	{632}
$\tfrac{3}{2}a:$	$b:$	$3c$	$3P\tfrac{3}{2}$	$3\text{-}\tfrac{3}{2}$	{231}

object is to its reflection in a mirror.* The *grade of symmetry* which any crystal form possesses is conditioned by the number of its planes of symmetry. For example, Figs. 28 and 29 represent two crystals, the one possessed of five planes of symmetry, the other of but one.

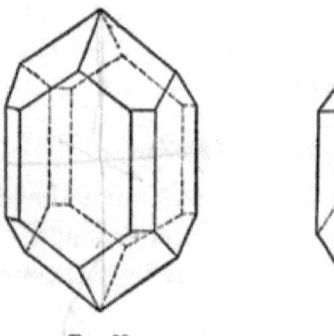
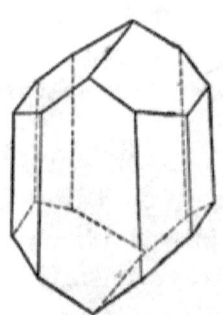

Fig. 28.   Fig. 29.

The law of symmetry may be formulated as follows: *All the faces of a crystal are grouped in accordance with certain definite planes of symmetry which are fixed in their position for the same crystal, and condition, not merely its external form, but, in an equal degree, the distribution of all of its internal physical properties.*

This law shows that symmetry is primarily a property of the internal molecular structure of crystals, and that, on this account, it is expressed in their outward form. This is the true explanation of its importance to crystallography.

In order to bring out clearly the symmetry of any

---

* More exactly we may say: two objects, or two halves of the same object, are *symmetrical* with reference to a plane placed between them, when from any point of one object a normal to this plane, prolonged by its own length on the opposite side of the plane, will meet the corresponding point of the other object.

crystal form, we must imagine it freed from all distortion and restored to its *ideal* proportions (see p. 14).

Planes of symmetry are of two kinds: those which contain two or more equivalent and interchangeable directions, and those which have no such directions. The first are called *principal*, and the others *secondary* planes of symmetry. This difference can be best illustrated in a concrete case. Fig. 30 represents a form two of whose axes, $a-a$ and $a'-a'$, are of equal length, while the third is of unequal length. According to our definition, the axial planes, $aa'aa'$ and $acac$, are both planes of symmetry; but, in the first, the directions $a-a$ and $a'-a'$ are equivalent and interchangeable, while, in the second plane, the directions $a-a$, $c-c$ are not interchangeable.

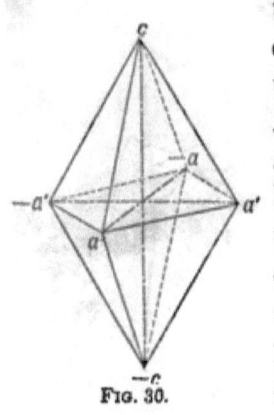

FIG. 30.

Again, the crystal form (Fig. 30) can be brought exactly into its original position by a revolution of less than 180° about the axis c; hence c is a *principal axis of symmetry*. It is, however, impossible to bring it into its original position by revolving it less than 180° about $a-a$ or $a'-a'$; hence these are called *secondary axes of symmetry*. Principal axes of symmetry are always normal to principal planes of symmetry, and secondary axes of symmetry, to secondary planes of symmetry.

Axes of symmetry are always chosen as the crystallographic axes, whenever there are three or more of them present. Moreover, when they are present, principal axes of symmetry are always preferred for this purpose to secondary axes. If only one principal axis is present, then two secondary axes are taken in addition. If only one axis of symmetry of any kind is present, then two other arbitrary directions are selected as crystallographic axes.

The Crystal Form. Thus far the term *form* has been employed in a somewhat loose sense. It has, however, in crystallography a very particular and technical meaning, which may be defined as *the sum of those planes whose presence is required by the symmetry of the crystal, when one of them is present*. In other words, a crystal form embraces all those planes which, irrespective of signs, can be represented by a single symbol. We shall in future employ the word *form* only in this special sense.

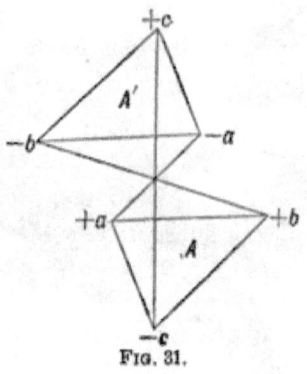
Fig. 31.

The number of planes composing a form depends upon the grade of symmetry and increases as the symmetry increases. For example, if a crystal is entirely without planes of symmetry, then the presence of any plane, A (Fig. 31), necessitates the occurrence of only its parallel plane, A'. In this case, therefore, a complete crystal form is composed of only two planes.

If, however, a single plane of symmetry ($WXYZ$) is present (Fig. 32), then each of the two planes, A and A', necessitates the occurrence of another plane (B and B') symmetrical to it. In this case the complete form consists of four similar planes.

Thus it can be seen that the number of planes

which go to make up a complete crystal form depends, in a way, upon the grade of symmetry of this form.

Crystal forms are divided into three *types*, according as their planes intersect one, two, or three of the axes of reference.

Pinacoids ($\pi i \nu \alpha \xi$, a board) are composed of planes parallel to two axes. They correspond in position to the faces of the cube.

Prisms are forms whose planes intersect two axes, while they are parallel to the third. When such forms are parallel to either of the two lateral axes, they are called *domes*.

Pyramids are forms whose planes cut all three axes.

Crystal forms are furthermore *closed* when their planes completely enclose space, and *open* when they do not. Those forms whose planes are all parallel to a single line are open forms. They cannot, of course, occur alone, but only in combination with other forms of the same grade of symmetry.

Crystal Combinations. Inasmuch as certain crystal forms do not by themselves enclose space, they cannot occur alone, and it is equally true that the occurrence alone of closed forms is rather the exception than the rule. The simultaneous occurrence on a single individual of two or more crystal forms is technically known as a *combination*.

The following points are of particular importance in regard to crystal combinations:

1. Only forms possessing the same grade of sym-

metry can combine. This is evident if we remember that the external form of a crystal is only an outward expression of its molecular structure; and that this must, of course, be the same throughout an entire individual.

2. When two or more forms combine, the axes of all must be coincident and possess the same relative, but not the same absolute lengths. This point is apt to present difficulties to a beginner. It must be remembered that it is the relative and not the absolute lengths of the axes which are essential in determining the position of any plane. A cube and octahedron, for example, whose axes are of equal length could not possibly combine, because the cube would entirely enclose the octahedron (Fig. 33). In order that these two forms may combine, the axes of the cube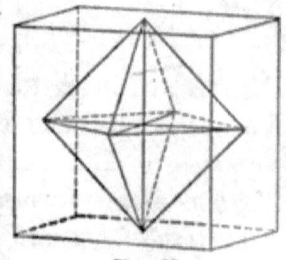

FIG. 33.

must be relatively the shorter. If they are only slightly so, the cube will appear as square truncations on the octahedral angles (Fig. 34). If, however, the cubic axes are much the shorter, the two forms will appear in equal development (Fig. 35), or the octahedron may form triangular truncations on the cubic angles (Fig. 36).

The relative development of the planes of different forms occurring in combination is an important factor in conditioning the habit of a crystal. This is also largely influenced by distortion and elongation, as explained on p. 14.

Certain terms are employed in describing the replacement of crystal edges and angles by the planes of other forms, which here need definition. The replacement of an edge or angle by a single plane is called a *truncation* (German, *Abstumpfung*); a replacement by two planes is called a *bevelment* (German, *Zuschärfung*).

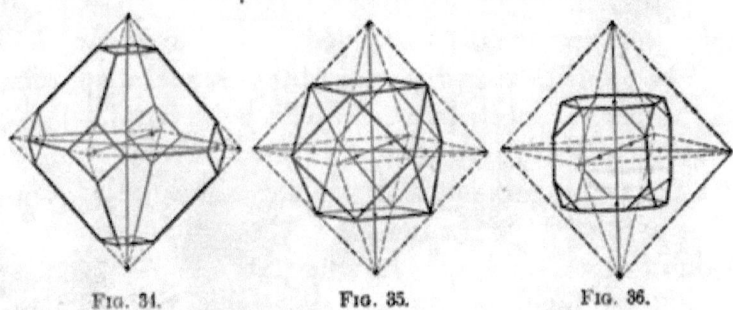

Fig. 34. Fig. 35. Fig. 36.

When a solid angle is replaced by more than two planes it is said to be *acuminated* or *blunted* (German, *Zuspitzung*). If a truncating plane makes equal angles with the planes on each side of it, it is said to be *symmetrical* (German, *gerade Abstumpfung*). If such angles are unequal, the truncation is *unsymmetrical* or *oblique* (German, *schiefe Abstumpfung*). Of the following figures, 37 and 38 are examples of symmetrical trunca-

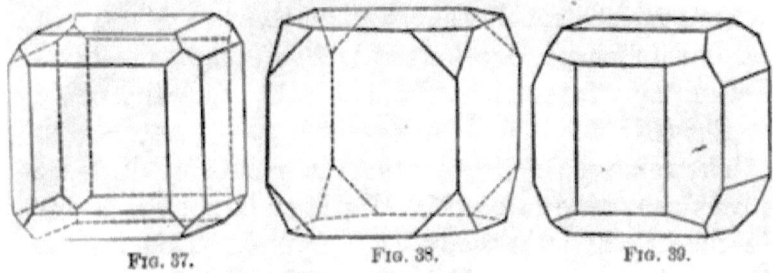

Fig. 37. Fig. 38. Fig. 39.

tions, the first of edges, the second of angles. Fig. 39 shows an unsymmetrical truncation of edges, while Figs. 40 and 41 represent bevelments.

It is a convenient fact to remember that the indices of any plane which symmetrically truncates an edge between two planes of the same form may be found by

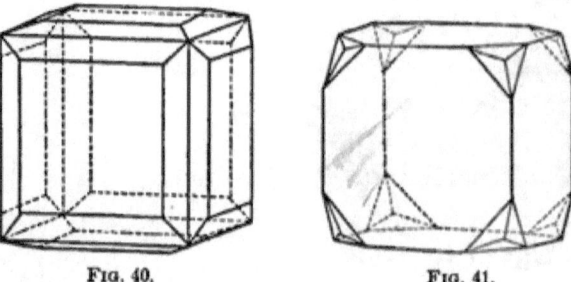

Fig. 40.   Fig. 41.

taking the algebraic sum of the indices of the two planes forming the edge truncated. Thus an edge between the planes 111 and 11$\bar{1}$ would be symmetrically truncated by the plane 110, an edge between 321 and 312 by 633 or 211, etc.

Independent Occurrence of Partial Crystal Forms. By the distortion of crystal forms due to irregular growth (p. 14), certain planes may be reduced in size to mere points and so disappear. Such an irregular absence of one or more planes belonging to a complete form is purely accidental and of no particular significance. It is known as *merohedrism* (μέρος, a part, and ἕδρα, face).

Hemihedrism and Tetartohedrism. In many other cases, however, crystal forms do occur, which, if we adhere to the definition of the *complete* form given on p. 35, must be regarded as *partial*. Such partial forms, on account of the regularity of their development and the frequency of their occurrence, are of the highest importance in crystallography; and to them must be accorded as full recognition as to those which fulfil all the requirements of symmetry. Their existence is best explained on the assumption that the planes composing *one half* or *one quarter* of certain complete forms are capable of occurrence, entirely independent of their other halves or quarters.

The theoretical consideration of all possible regular arrangements of points in space which satisfy the conditions of crystal structure (p. 6) shows that these also include partial forms, similar to those observed in nature.

In contrast to such partial forms, the completely symmetrical crystal is termed *holohedral* (ὅλος, whole, and ἕδρα, face), or a *holohedron*; while the half forms are called *hemihedral* (ἥμι, half, and εδρα, face) or *hemihedrons*, and the quarter forms are known as *tetartohedral* (τέταρτος, quarter, and ἕδρα, face) or *tetartohedrons*.

We may imagine any hemihedral or tetartohedral form as produced by the suppression of a certain half or three quarters of the planes composing the complete or holohedral form, and the extension of those planes which remain until they meet. Let us, for instance, suppose that the white planes of the inside figure (Fig. 42) are suppressed, and that the shaded planes are extended to intersection; they will then produce the outside figure, which is the hemihedron corresponding to the interior holohedron.

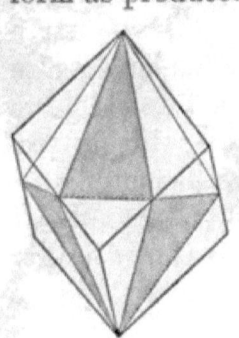

Fig. 42.

Hemihedral and tetartohedral forms cannot, however, be produced by the selection and extension of any arbitrary half of the planes belonging to the corresponding holohedral form. On the contrary, the planes capable of producing partial forms must be selected in accordance with certain definite conditions. These may be stated as follows: *If the crystal be imagined as free from all distortion, then only such planes of the complete form can survive to produce a hemihedron or tetartohedron as will, after their extension, intersect the extremities of all equivalent axes of symmetry in the same number, under equal angles and at the same distance from the centre.*

Figs. 43 and 44 show how the halves of the planes

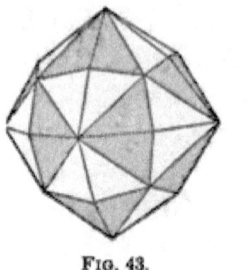
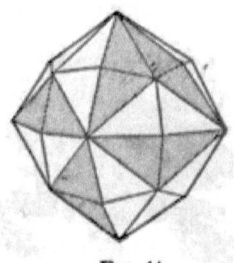

FIG. 43.   FIG. 44.

of a holohedron may be selected so as in the first instance to fulfil, and in the second not to fulfil, the above conditions. For in the first case the vertical axis is cut at its extremities by pairs of shaded planes, just as the lateral axes are; while, in the second case, the vertical axis is still cut by pairs of shaded planes, but the lateral axes by single ones, which alternate with white planes.

Every complete crystal form is bounded by pairs of parallel planes (p. 18). Such a form may become hemihedral (1) by losing one half of its pairs of planes, the other pairs remaining intact; or (2) by losing one plane from each of its pairs, so that no plane of the resulting hemihedron has another parallel to it. The first method produces what is called *parallel-face hemihedrism* (Fig. 45), and the second what is known as *inclined-face hemihedrism* (Fig. 46).

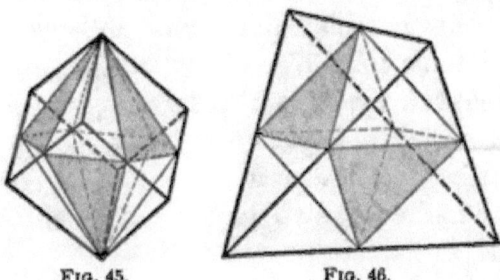

FIG. 45.   FIG. 46.

Inasmuch as all crystal forms are the external expressions of an internal molecular structure, the law of combinations (p. 36) must hold good not merely for holohedral, but also for hemihedral and tetartohedral crystals. *Only forms belonging to the same kind of hemihedrism or tetartohedrism can combine.* Apparent exceptions to this law are caused by the fact that certain holohedral forms are incapable of producing hemihedrons which are geometrically different from themselves. This will be more fully explained in the succeeding chapters.

Hemimorphism. One half of the planes bounding a holohedron sometimes occur independently of the other half, where the selection of faces cannot be brought within the above-given conditions of hemihedrism. Instead of the new form having one half of the planes similarly grouped about either extremity of an axis of symmetry, we may have all of the holohedral planes at one extremity of such an axis, and none of them at the other. This mode of development is called *hemimorphism* (ἡμι, half, μόρφη, form), and the axis with reference to which the planes are grouped is known as the *hemimorphic axis*.

Hemimorphic forms cannot of themselves enclose space. They must, therefore, always occur in combination with other hemimorphic forms, as is shown in the case of calamine (Fig. 47). That this development is the direct result of molecular structure is shown both by the hemimorphic indentations produced when the crystal planes are etched, and by certain peculiar physical properties, most prominent among which is pyro-electricity. When hemimorphic crystals are heated, they give off positive electricity at one end (*analogue pole*), and negative at the other (*antilogue pole*). When the temperature begins to fall, this order is reversed.

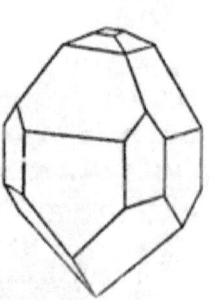

FIG. 47.

Symmetry of a Crystal Plane. The symmetry of a crystal plane is conditioned by the number of planes of symmetry to which it is normal. Thus a face perpendicular to one plane of symmetry is called *monosymmetric;* one perpendicular to two planes of symmetry, *disymmetric;* etc.

The Crystal Systems. It can be proved in a variety of ways that all the complete or holohedral crystal forms that are possible must possess one of six grades of symmetry. Symmetry, therefore, furnishes a valuable means of classifying the complete forms into six groups, called *Crystal Systems*. From the symmetry of the holohedral forms the crystallographic axes, to which their planes are referred, are deduced, as already stated on p. 34.

A definition of the crystal systems, however, which is based entirely upon symmetry cannot be wholly satisfactory, since it strictly excludes the partial crystal forms (hemihedral, tetartohedral, and hemimorphic), whose grade of symmetry is always less than that of their corresponding holohedrons. These partial forms are nevertheless referable to the same crystallographic axes as the complete forms from which they are derived, and hence the crystal systems must be defined in terms of both their symmetry and axes: *A system is the sum of all the possible crystal forms whose planes can be referred to the same kind of axes; or the sum of all possible* HOLOHEDRAL *forms which possess the same grade of symmetry.*

According to this definition, we may characterize the six crystal systems as follows:

CLASS I. (Isometric.) 1. All forms referable to three axes of equal length which intersect at angles of 90°. Holohedral forms are possessed of three principal planes of symmetry at right angles to one another (giving the three rectangular axes); and six secondary planes of symmetry, which bisect each of the angles between the principal planes. . . Isometric System.

CLASS II. (Isodimetric.) All forms referable to one principal or vertical axis, which is perpendicular to, and different in length from the lateral axes. One principal plane of symmetry, giving the principal axis.

2. Number of equal lateral axes two, intersecting the principal axis and each other at angles of 90°. Number of secondary planes of symmetry for holohedral forms four, which are all perpendicular to the principal plane of symmetry, and inclined to each other at angles of 45°, 90°, and 135°. . Tetragonal System.

3. Number of equal lateral axes three, intersecting the principal axis at angles of 90°, and each other at angles of 60°. Number of secondary planes of symmetry for holohedral forms six, all at right angles to the principal plane of symmetry, and inclined to each other at angles of 30°, 60°, 90°, 120°, and 150°.

Hexagonal System.

CLASS III. (Anisometric.) No principal axis or plane of symmetry.

4. All forms referable to three axes of unequal length intersecting at right angles. Holohedral forms possessed of three secondary planes of symmetry at right angles. . . . . Orthorhombic System.

5. All forms referable to three axes of unequal length, two of which intersect at an oblique angle, while they are both perpendicular to the third. One secondary plane of symmetry. . . . Monoclinic System.

6. All forms referable to three axes of unequal length, all oblique to one another. No plane or axis of symmetry. . . . . Triclinic System.

We may now proceed to the description of the particular types of crystal forms which compose each of these six systems.

CHAPTER III.

THE ISOMETRIC SYSTEM.*

HOLOHEDRAL DIVISION.

Symmetry. The special characteristics of a crystal system may be advantageously deduced from the symmetry of its holohedral forms, for this belongs primarily to the molecular structure of the crystals themselves.

According to the definition of the isometric system given in the preceding chapter, its complete forms possess the highest grade of symmetry which is consistent with the law of rational indices. This is nine-fold, and is distributed about three principal planes of symmetry at right angles to one another, the angles between which are bisected by six other secondary planes of symmetry. The position of these planes of symmetry is shown in the accompanying figure (No. 48), where the three principal planes are shaded and indicated by Roman numerals, while the secondary planes are white, and are numbered 1, 2, 3, 4, 5, 6.

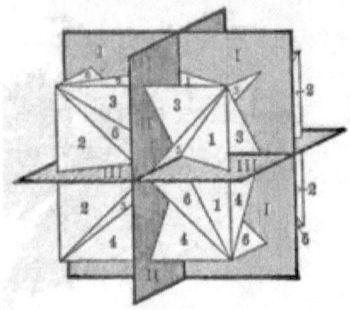

FIG. 48.

The symmetry of partial forms in the isometric system is necessarily less than that of the complete

---

* Also called the *regular*, *tesseral*, *tessular*, or *cubic* system.

forms, and its character will be fully explained beyond, as each of the hemihedral and tetartohedral divisions is considered.

Axes. A still more comprehensive definition of the isometric system, inasmuch as it includes both partial and complete forms, is that it is *the sum of all crystal forms whose planes are referable to three axes of equal length which intersect at right angles*. These axes are directly deducible from the holohedral isometric symmetry, because they represent the three principal axes of symmetry; and, as has been already stated (p. 34), principal axes of symmetry are employed as crystallographic axes whenever they are present. Moreover, because these axes lie, two and two, in the principal planes of symmetry, they must be not merely *rectangular and of equal length, but also all interchangeable; i.e., whatever is true of one must be true also of the other two.**

The Fundamental Form. The starting-point for any series of planes which is referable to the same set of rectangular crystallographic axes, is a form which cuts all of the axes at their unit length. This is called the fundamental or ground-form for the series. In the isometric system all of the axes have the same length, and its

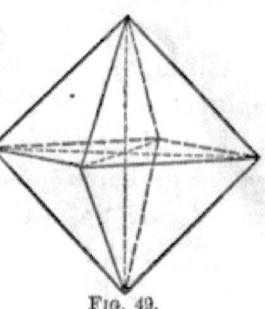

FIG. 49.

---

* Two other sets of axes are of use in the isometric system. One is the set of intersection-lines between the principal and secondary planes of symmetry; and the other the intersection-lines of the secondary planes of symmetry with each other. The first are normal to the faces of the rhombic dodecahedron and are called the *digonal;* the second are normal to the faces of the octahedron and are called the *trigonal axes.*

ground-form must therefore be one whose planes cut all the axes at the same distance from the centre. This is the *regular octahedron*, whose eight equilateral sides intersect at angles of 109° 28′ 16.4″ (Fig. 49). The parameters and indices of this form are all alike unity. Modified by signs indicating particular octants, as explained on p. 31, the indices of the octahedron are:

(Above)   111,   1$\bar{1}$1, $\bar{1}\bar{1}$1, $\bar{1}$11.
(Below)   11$\bar{1}$,   1$\bar{1}\bar{1}$, $\bar{1}\bar{1}\bar{1}$, $\bar{1}$1$\bar{1}$.

Derivation of the Types of Holohedral Forms possible in the Isometric System. Before attempting to describe the isometric crystal forms, it will be well to discover how many types of such forms are possible; i.e., how many different kinds of forms possess the complete isometric symmetry and have their planes referable to three equal and rectangular axes. To do this, we may first find what is the most general form consistent with these conditions, and then deduce from this all possible special cases, by giving definite or limiting values to one or both of its parameters.

The most general symbol for any plane referred to a set of three perpendicular axes, expressed in the parameter notation of Weiss, is, as was explained on p. 26, $na : b : mc$. Since, however, in the isometric system all three axes are equal and interchangeable, they are represented by the same letter, a, and this general formula therefore becomes $na : a : ma$.

Of this symbol, as it stands, six, and only six, permutations are possible, as follows:

$na : a : ma;$ $a : na : ma;$ $ma : a : na;$
$na : ma : a;$ $a : ma : na;$ $ma : na : a.$

Inasmuch as the signs of these six symbols are

THE ISOMETRIC SYSTEM.

throughout the same, all the planes which they represent must belong to a single octant (p. 31). The three principal planes of symmetry of course require the repetition of the same group of six planes in each of the eight octants into which they divide space; and hence the most general isometric form must be bounded by forty-eight planes.

The same conclusion may be reached by writing the general expression for the indices of a form, $\{hkl\}$, in every order and with every possible combination of signs. In this way we obtain the following forty-eight symbols, each of which stands for one particular plane of the most general isometric form:

Four upper octants.				Four lower octants.			
hkl	$h\bar{k}l$	$\bar{h}\bar{k}l$	$\bar{h}kl$	$hk\bar{l}$	$h\bar{k}\bar{l}$	$\bar{h}\bar{k}\bar{l}$	$\bar{h}k\bar{l}$
hlk	$h\bar{l}k$	$\bar{h}\bar{l}k$	$\bar{h}lk$	$hl\bar{k}$	$h\bar{l}\bar{k}$	$\bar{h}\bar{l}\bar{k}$	$\bar{h}l\bar{k}$
khl	$k\bar{h}l$	$\bar{k}\bar{h}l$	$\bar{k}hl$	$kh\bar{l}$	$k\bar{h}\bar{l}$	$\bar{k}\bar{h}\bar{l}$	$\bar{k}h\bar{l}$
klh	$k\bar{l}h$	$\bar{k}\bar{l}h$	$\bar{k}lh$	$kl\bar{h}$	$k\bar{l}\bar{h}$	$\bar{k}\bar{l}\bar{h}$	$\bar{k}l\bar{h}$
lhk	$l\bar{h}k$	$\bar{l}\bar{h}k$	$\bar{l}hk$	$lh\bar{k}$	$l\bar{h}\bar{k}$	$\bar{l}\bar{h}\bar{k}$	$\bar{l}h\bar{k}$
lkh	$l\bar{k}h$	$\bar{l}\bar{k}h$	$\bar{l}kh$	$lk\bar{h}$	$l\bar{k}\bar{h}$	$\bar{l}\bar{k}\bar{h}$	$\bar{l}k\bar{h}$

Each vertical column here represents the planes of a single octant. The order of the octants, commencing with the upper, right, front one, is around the upper half of the crystal from right to left, and then, in a similar way, around the lower half.

The number of planes belonging to the most general isometric form may be arrived at in still another way. The nine planes of symmetry (see Fig. 48) divide space into forty-eight equal triangular secants. The presence of any plane in one of these secants, oblique to all the planes of symmetry, necessitates another plane, similarly inclined, in each of the other forty-seven secants. Hence forty-eight is the number of planes bounding the most general form.

The parameters, m and n, in the most general symbol stand for any rational quantity greater than one and less than infinity. The reason why they are never made less than one is because in the isometric system all of the axes are equal and interchangeable; hence any one of the axial intercepts for any plane may be assumed as unity, and it is customary to make the shortest of them unity. Now all the other types of holohedral forms possible in the isometric system may be deduced as special cases of the most general type, $na : a : ma$, by giving limiting values to one or both of its parameters. These limiting values are three in number, viz., the smallest possible value, unity; the largest possible value, infinity; and a value for one parameter equal to that of the other. Since special cases may be produced by limiting either one or both of the variable parameters, we find that seven and only seven distinct types of isometric holohedrons are possible, which fall naturally into the three following classes:

Class I. Forms with *two* variable parameters.

1. $m > n$, general symbol becomes $na : a : ma$.

Class II. Forms with *one* variable parameter.

2. m or $n = \infty$, general symbol becomes
$$\infty a : a : ma.$$

3. m or $n = 1$, general symbol becomes $a : a : ma$.

4. $m = n$, general symbol becomes $ma : a : ma$.

Class III. Forms with *no* variable parameter.

5. $m = 1$, $n = \infty$, general symbol becomes
$$\infty a : a : a.$$

6. m and $n = \infty$, general symbol becomes
$$\infty a : a : \infty a.$$
7. m and $n = 1$, general symbol becomes $a : a : a$.

If we employ the indices where $h > k > l$, the symbols corresponding to these seven types become $\{hkl\}$, $\{hk0\}$, $\{hll\}$, $\{hkk\}$, $\{110\}$, $\{100\}$, and $\{111\}$.

The particular characters of these seven isometric form-types we shall now proceed to consider in succession.

The Hexoctahedron. We shall best be able to appreciate the nature of the most general or forty-eight-

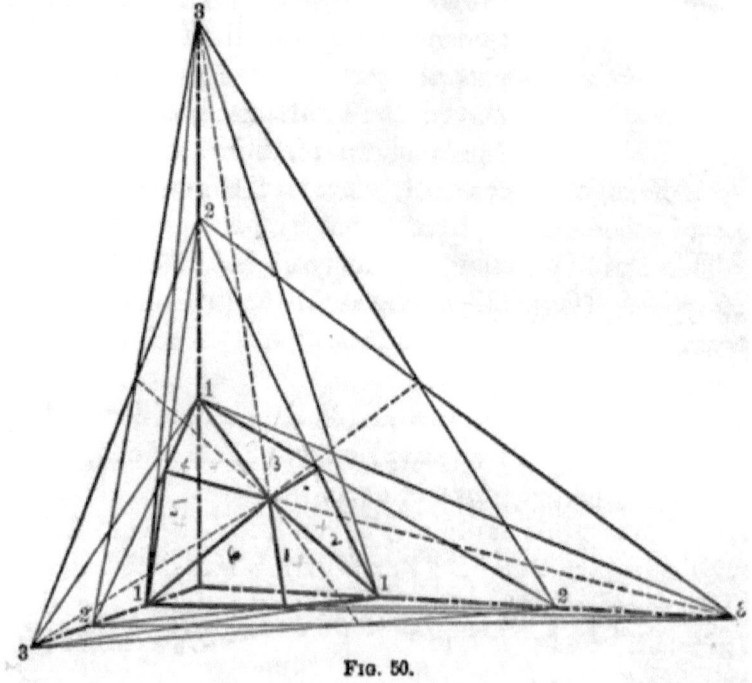

Fig. 50.

sided form if we construct upon three equal axes the planes which occupy a single octant. For this purpose we may assume the definite values $n = 2$ and

$m = 3$. Then the possible permutations of the symbol become:

$2a : a : 3a$; $a : 2a : 3a$; $3a : a : 2a$;
$2a : 3a : a$; $a : 3a : 2a$; $3a : 2a : a$.

Each expression represents one plane of the same octant, which may be independently constructed on the axes. The intersections of the six planes thus obtained give the assemblage represented in Fig. 50. The principal planes of symmetry now require the occurrence of a similar group in each of the other seven octants, and the result is the solid figure shown in Fig. 51. Each plane of the fundamental form is here replaced by a hexagonal pyramid, whence the name for this form is the *hexoctahedron*.

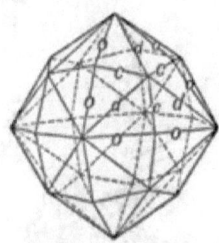

Fig. 51.

This form is bounded by forty-eight similar scalene triangles. Its solid angles are of three kinds: six octahedral, at the extremities of the principal axes; eight hexahedral, at the extremities of the trigonal axes; and twelve tetrahedral, at the extremities of the digonal axes. There are also three kinds of combination-edges, twenty-four of each. Of these the shorter, for reasons to be explained beyond, may be called the *cubic* edges (c, Fig. 51); those of intermediate length, the *octahedral* edges (o, Fig. 51); and the longer, the *dodecahedral* edges (d, Fig. 51). Naumann's symbol for this form is $m\,O\,n$; Dana's, **m-n**.

The Tetrahexahedron. If one of the two variable parameters be given the limiting value infinity, the general formula becomes $\infty a : a : ma$. This represents a figure bounded by twenty-four sides, each of which is parallel to one of the principal axes. It is therefore a combination of planes of the prismatic type (p. 36). We may make a construction of the planes in one octant, as was done in the last case, assuming for m the value 3. The possible permutations then become:

$\infty a : \ a : 3a;$ $\quad a : \infty a : 3a;$ $\quad 3a : \infty a : \ a;$
$\infty a : 3a : \ a;$ $\quad a : 3a : \infty a;$ $\quad 3a : \ a : \infty a.$

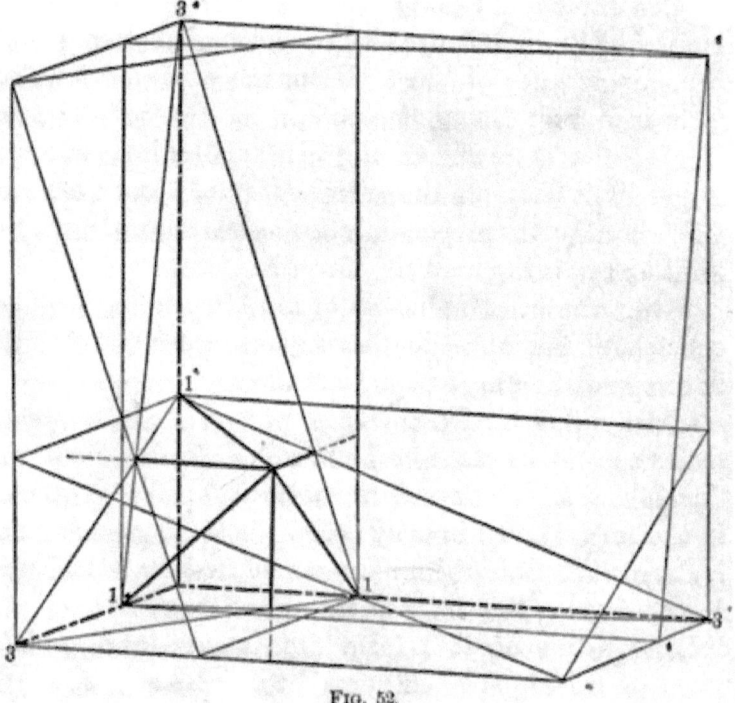

Fig. 52.

The resulting group of planes is not, at first glance, very different from that obtained before; but here each plane is normal to a principal plane of symmetry, and there-

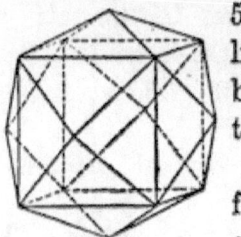

Fig. 53.

fore belongs equally to two contiguous octants (Fig. 52). The complete form (Fig. 53) is like a cube whose faces are replaced by a quadratic pyramid, and it is therefore called the *tetrahexahedron*.

The planes of this form are twenty-four similar isosceles triangles. Its solid angles are of two kinds: six tetrahedral, at the ends of the principal axes; and eight hexahedral, at the ends of the trigonal axes. Twelve cubic and twenty-four dodecahedral edges remain, but the octahedral edges have disappeared by becoming angles of 180°. Naumann's symbol for this form is $\infty O n$; Dana's is $i\text{-}n$.

The Trisoctahedron. Giving n its smaller instead of its larger limit changes the general formula to $a : a : ma$. Of this only three permutations are possible, which indicates that the resulting form has but three planes in an octant. A construction similar to those given in the preceding cases can be readily made by the student. This produces a group of planes, which, when developed for each octant, results in the form shown in Fig. 54. It resembles an octahedron with each of its planes replaced by a triangular pyramid, whence its name—the *trisoctahedron*.

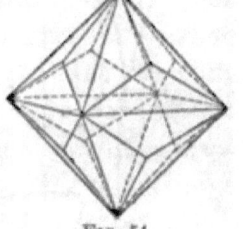

Fig. 54.

The planes of this form are twenty-four similar isosceles triangles, each normal to a plane of symmetry, and hence monosymmetric (p. 43). Its solid angles are of two kinds: six octahedral, at the ends of the principal axes; and eight trihedral, at the ends of the trigonal axes. There are twelve octahedral and twenty-four dodecahedral edges, while the cubic edges have disappeared, like the octahedral edges in the tetrahexahedron. Naumann's symbol for this form is mO; Dana designates it by m, and calls it the *trigonal trisoctahedron*.

The Icositetrahedron. When $n = m$, the general formula becomes $ma : a : ma$, of which there are again but three permutations possible. A construction like the others, with some definite value assumed for m, yields three planes in each octant, and a completed figure like that shown in Fig. 55. In allusion to its being twenty-four-sided, it is usually called the *icositetrahedron*.*

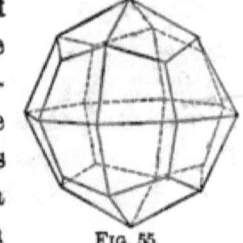

FIG. 55.

The planes of this form are similar trapeziums. Its solid angles are of three kinds: six tetrahedral at the ends of the principal axes; twelve tetrahedral at the ends of the digonal axes; and eight trihedral at the ends of the trigonal axes. There are twenty-four octahedral and twenty-four cubic edges, while the dodecahedral edges have disappeared. Naumann's symbol for this form is mOm; Dana's is $m\text{-}m$.

The Rhombic Dodecahedron. If one parameter is given its largest, and the other its smallest limiting value, the general formula becomes, $a : a : \infty a$. This is capable of three permutations, which locates three planes in each octant; but the sign of infinity shows that each plane is parallel to an axis, and therefore common to two contiguous octants. The result must

---

* It is also known as the *trapezohedron*, the *leucitohedron*, and *tetragonal trisoctahedron*.

be a form bounded by twelve planes, which the construction shows are similar rhombs (Fig. 56). This form is called the *rhombic dodecahedron*. It has six acute tetrahedral angles at the extremities of the principal axes, and eight obtuse trihedral angles at the ends of the trigonal axes. Its edges are twenty-four in number, and are all of one sort (*dodecahedral*), enclosing angles of 120°. Each of its faces is normal to two planes of symmetry, and therefore disymmetric. Naumann's symbol for this form is ∞ O; Dana's is I.

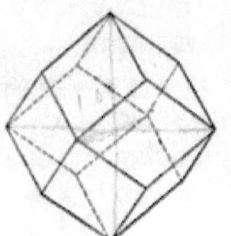

Fig. 56.

The Hexahedron (*Cube*). If both parameters reach their maximum limit, the general formula becomes ∞ a : a : ∞ a, also capable of three permutations. Hence there are three planes in each octant, but the two signs of infinity show that every plane is parallel to two axes, and hence is common to four contiguous octants. The result is a six-sided figure, which is therefore called the *hexahedron* (Fig. 57). Its sides are squares, normal to four planes of symmetry, and quadrisymmetric. It has eight similar trihedral angles at the ends of the trigonal axes, and twelve similar edges (*cubic*) enclosing angles of 90°. Naumann's symbol for the cube is ∞ O ∞; Dana's is i-i.

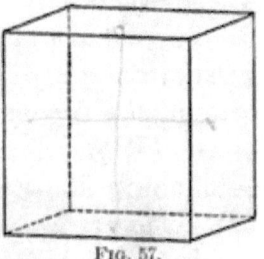

Fig. 57.

The Octahedron. The simplest form to which the most general isometric parameter symbol can be reduced is a : a : a, which represents one plane in each of the eight octants, intersecting all of the axes at unit distance from the centre. This produces the regular *octahedron*, or fundamental form of the system (see p. 47, Fig. 49). Its faces are all equilateral triangles; its six solid angles all similar and tetrahedral, situated at the extremities of the axes; and its edges are all similar (*octahedral*). Naumann abbreviates the parameter symbol of this form into its initial letter, O; while Dana designates it by the figure 1.

Relations of the Seven Isometric Holohedrons to each other—Limiting Forms. We have seen (p. 50) that the six simpler types of isometric holohedrons may be regarded as special cases of the most general type or hexoctahedron. Of course those types whose symbols contain one or more variables may be represented by a variety of forms, which are alike in the number, distribution, and symmetry of their faces, but which differ in their corresponding interfacial angles. Thus $2O2$, $3O3$, and $5O5$ all represent icositetrahedrons, although their corresponding angles are not identical. On the other hand, only one representative is possible of those types whose symbols contain no variable parameter; and hence all cubes, octahedrons, and rhombic dodecahedrons must be alike. These are therefore called *fixed* forms.

In proportion as the parameters in any symbol approach their limiting values, so do the forms which they represent approach nearer to their limiting forms. Plate I. is arranged to show the relations of the isometric holohedrons in this respect. The three fixed or unvariable forms occupy the corners of the triangular diagram, since they are the final limits between which the other types vary. The colors of their edges —octahedral (black), cubic (blue), and dodecahedral (red)—are retained in all the other forms. Between each pair of fixed forms oscillates one of the three types of twenty-four-sided figures which have but a single variable parameter in their symbols.

Let us, as an illustration, consider the case of the trisoctahedron. The number of trisoctahedrons theoretically possible is infinite, because m may be given an infinite number of rational values; practically the number is small because, on natural crystals, m is generally some small whole number. Such a sequence of trisoctahedrons may be represented by the following symbols: $\frac{6}{5}O, \frac{4}{3}O, \frac{3}{2}O, 2O, \frac{5}{2}O, 3O, 4O, 6O, 12O, 18O$, etc. As the value of m decreases, the more obtuse do the dodecahedral (red) edges become; until when m reaches its lower limit, 1, the angles disappear by becoming $= 180°$, and the form merges into the octahedron—one of its limiting forms. As the value of m increases, the more acute the dodecahedral edges of the trisoctahedron and the more obtuse its octahedral edges become. When m reaches its upper limit, ∞, the former have become $120°$ and the latter $180°$, i.e., the trisoctahedron has graded into the rhombic dodecahedron, the other of its limiting forms.

Exactly the same gradations may be traced for the sequence of icositetrahedrons between its limiting forms, the octahedron and the cube; and for the possible tetrahexahedrons, between the cube and dodecahedron.

Each type of the twenty-four-sided figures possesses the two sorts of edges belonging to its own limiting forms. Each can oscillate backward or forward along a *single* direction, because it has but a single variable in its symbol.

THE ISOMETRIC SYSTEM. 59

The variations of the hexoctahedral type, with two variable quantities in its symbol, $na : a : ma$, are more complex. The value of one parameter may be changed, while the other remains fixed; or both may be changed simultaneously. Thus sequences of hexoctahedrons may be deduced, oscillating along different lines, and between different pairs of limiting forms.

Suppose we start with the hexoctahedron $4O2$, we may vary but one parameter in three ways: (1) By increasing the greater: $4O2, 5O2, 6O2, 12O2, 18O2$, etc., to $\infty O2$, the tetrahexahedron, one limiting form. (2) By decreasing the greater to the limit of the less: $4O2, \frac{7}{2}O2, 3O2, \frac{5}{2}O2$, etc., to $2O2$, the icositetrahedron, another limiting form. (3) By decreasing the less, $4O2, 4O\frac{7}{4}, 4O\frac{3}{2}, 4O\frac{5}{4}$, etc., to $4O$, the trisoctahedron, a third limiting form.

Both parameters may also be varied simultaneously in three ways: (1) Both may be increased: $4O2, 5O3, 6O4, 8O6$, etc., to $\infty O \infty$, the cube. (2) Both may be diminished: $4O2, \frac{7}{2}O\frac{7}{4}, 3O\frac{3}{2}, \frac{5}{2}O\frac{5}{4}$, to O, the octahedron. (3) One may be increased, while the other is diminished: $4O2, 5O\frac{7}{4}, 6O\frac{3}{2}, 7O\frac{5}{4}$ to ∞O, the rhombic dodecahedron.

Thus we see that all the other holohedral forms of the isometric system are not merely special cases of the hexoctahedron, but that they are as well its limiting forms in different directions.

The Holohedral Isometric Forms in Combination. Although all isometric forms completely enclose space, and are therefore capable of independent occurrence, still most isometric crystals exhibit the occurrence of two or more forms together. These combinations are far too manifold for special description. Their acquaintance must be made by practice with models and natural crystals.

The figures on Plate II. are arranged to show each of the isometric holohedrons in two different combinations with all the others. The simple forms run in a diagonal through the plate. The combinations above this line show the simpler, and those below it the more complex forms predominating.*

Only a few points relating to these combinations can be specified here. We notice that the cube and octahedron mutually truncate each other's solid angles, while the edges of both forms are replaced by the

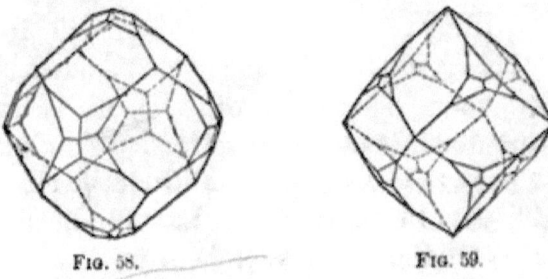

FIG. 58.   FIG. 59.

planes of the rhombic dodecahedron. In combinations of fixed with variable forms, or of variable forms with each other, the particular mode of replacement depends on the values of the parameters. For instance, the faces of the icositetrahedron truncate the edges of the rhombic dodecahedron only when the parameters m, of the former, are equal to *two* (Pl. II. fig. 18). If these parameters are greater than two, its faces replace the tetrahedral; if less than two, the trihedral angles of the dodecahedron (Figs. 58 and 59). Similar relations,

---

* From Ulrich's Krystallographische Figurtafeln. 4°; Hannover, 1884.

which will be readily understood, are indicated by dotted lines on Figs. 18, 26, 27, 39 and 46, of Plate II.

Combinations of three or more forms are quite as common as those of two. A few examples of such complex combinations are given in the following figures:

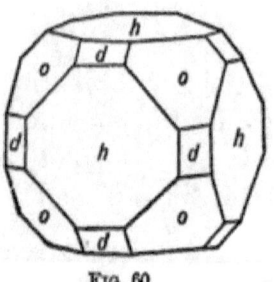
Fig. 60.

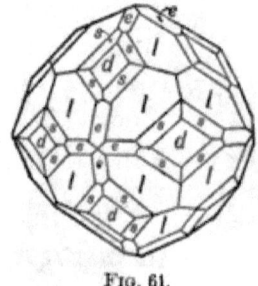
Fig. 61.

Fig. 60 (galena) shows the cube (h), the octahedron (o), and the dodecahedron (d).

Fig. 61 (garnet) shows the dodecahedron, the icositetrahedron, $2O2$, $\{211\}$ (l), the tetrahexahedron, $\infty O2$, $\{210\}$ (e), and the hexoctahedron, $3O\frac{3}{2}$, $\{231\}$ (s).

Fig. 62 (amalgam) shows the cube, the dodecahedron, the tetrahexahedron, $\infty O3$, $\{310\}$ (f), and the icositetrahedron, $2O2$, $\{211\}$ (l).

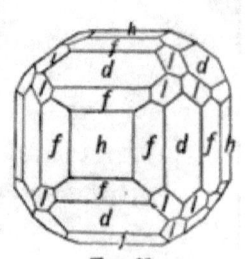
Fig. 62.

The following isometric substances may be mentioned as examples of holohedral crystallization: the metals copper, silver, gold, and platinum; amalgam (HgAg); the sulphides of lead (galena) and of silver (argentite); the chlorides of sodium (halite) and of silver (cerargyrite); calcium fluoride (fluor-spar); spinel, garnet, microlite, sodalite, nosean, leucite, and analcite.

HEMIHEDRAL DIVISION OF THE ISOMETRIC SYSTEM.

Kinds of Hemihedrism. As has been explained in the preceding chapter, one half of the planes composing a complete form may occur on crystals of certain substances independently of the other half. Such partial forms are, however, only possible when the planes which compose them are selected from the corresponding complete form in accordance with certain fixed conditions, viz.: they must intersect the extremities of all equivalent axes of symmetry in the same number, under equal angles and at equal distances from the centre (p. 41).

In order to discover how many different kinds of hemihedral forms are possible in the isometric system, we may imagine one half of the planes bounding the most general form—the hexoctahedron—to be selected in every conceivable way, and then note which of these ways satisfy the above conditions. Such a method of procedure shows that, although we can choose one

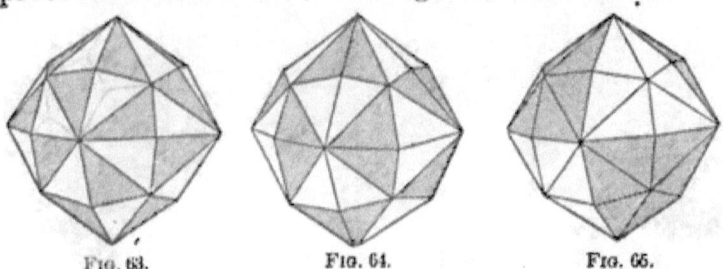

FIG. 63.   FIG. 64.   FIG. 65.

half of the forty-eight planes of the most general holohedron in a great number of different ways, only *three* of them yield crystallographically possible partial forms. These three methods of selection, each capable of producing a different kind of isometric hemihedrism, are illustrated in the three figures, 63–65. The first

THE ISOMETRIC SYSTEM. 63

(Fig. 63) is a selection by alternate planes; the second (Fig. 64) is by pairs of planes which intersect in the principal planes of symmetry, or in the octahedral edges; and the third (Fig. 65) is a selection by alternate octants. If we imagine either the white or shaded planes of these figures to disappear, and the others to be extended until they intersect, three new forms will result, all of which satisfy the conditions of hemihedrism, and which occur on natural crystals. Thus three distinct kinds of isometric hemihedrism are possible. These are called, for reasons which will appear as each in turn is considered:

1. Gyroidal, or plagiohedral hemihedrism;
2. Pentagonal, or parallel-face hemihedrism;
3. Tetrahedral, or inclined-face hemihedrism.

Gyroidal Hemihedrism. If the alternate planes of the most general isometric form—the hexoctahedron—are

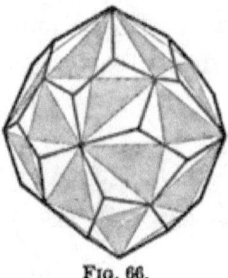
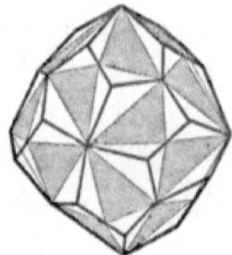

FIG. 66.　　　　　FIG. 67.

extended until they intersect, a new form will result bounded by twenty-four similar but unsymmetrical pentagons. Two such forms must be derivable from every hexoctahedron, one produced from one set and the other from the other set of alternate planes (Figs. 66 and 67). They are called *pentagonal icositetrahedrons*, and are distinguished as right- and left-handed according as they contain the right or left top plane of the front, upper octant. Their symbols, in the notations of Naumann and Miller, are

$$\frac{mOn}{2}r, \quad \gamma\{lkh\} \quad \text{and} \quad \frac{mOn}{2}l, \quad \gamma\{klh\},$$

where $h > k > l$, as in the case of holohedral forms (p. 51).

Apparently Holohedral Hemihedrons. Since all the other six isometric holohedrons may be regarded as special cases of the hexoctahedron (p. 50), we may consider them all as forty-eight-sided figures, certain of whose planes intersect at angles of 180°. From this point of view, the faces of the cube are composed of eight planes; those of the octahedron of six; those of the dodecahedron of four; etc. Now if the above method of selection be applied to the forty-eight planes, by which every isometric holohedron may be considered as bounded, it is evident that in every case, except the most general, the surviving planes will, by their extension, reproduce the form without geometrical change. This must be so, because every face of the six more special forms is made up of at least two contiguous planes of the general form, intersecting at angles of 180°; and hence either half of any face is enough to reproduce it. This may be seen from the six following figures, whose planes are shaded in accordance with the alternate method of selection. Thus only one geometrically new form-type is possible by gyroidal hemihedrism, but other apparently holohedral forms combined with it must be regarded as just as truly hemihedral. Their essential character consists in their molecular structure, and this must be the same throughout the same crystal individual. If a given molecular structure can produce *any* form which is geometrically hemihedral, this is a proof that *all* the forms produced by the same structure must be equally hemihedral, whether they have a different

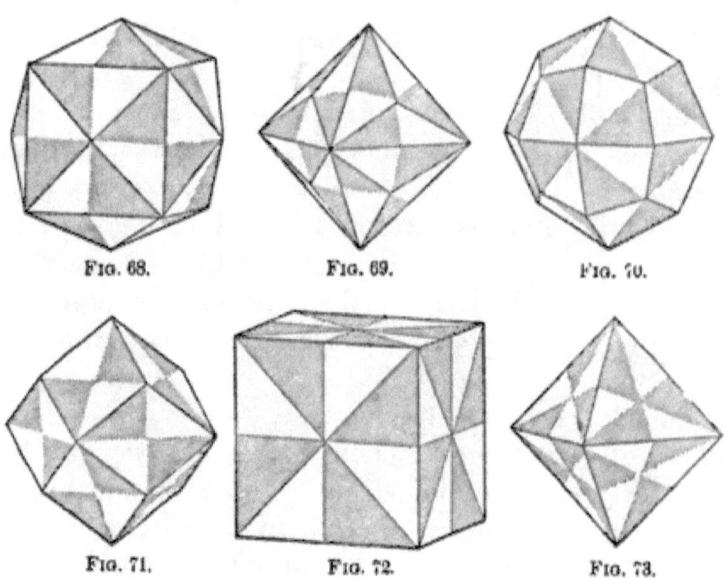

Fig. 68.   Fig. 69.   Fig. 70.
Fig. 71.   Fig. 72.   Fig. 73.

shape from holohedral forms or not. Thus we may have two cubes or octahedrons which are outwardly alike but in reality different; and their difference can, in many cases, be demonstrated by their physical behavior or by etching.

Symmetry of Gyroidal Forms. An inspection of the only new gyroidal hemihedron—the pentagonal icositetrahedron—shows that it possesses no plane of symmetry whatever. All of the nine isometric planes of symmetry disappear by the alternate method of selection. Nevertheless, the faces of this form are referable to three equal and rectangular axes, and it is therefore to be reckoned in the isometric system. Although the right- and left-handed pentagonal icositetrahedrons themselves are without symmetry, they are still symmetrical with reference to each other. Symmetrical forms, which are themselves devoid of planes of symmetry and cannot therefore be brought by any revolution into exactly the same position, are called in crystallography *enantiomorphous* (from ἐνάντιος, opposite, and μόρφη, form).

In spite of their apparently holohedral shape, all the other gyroidal forms of the isometric system must also be considered as devoid of symmetry, because they are produced by an asymmetric molecular arrangement.

Gyroidal hemihedrism is not of common occurrence, as it has, up to the present time, been observed on the crystals of only three substances: cuprite (Cu_2O), sylvine (KCl), and sal-ammoniac (NH_4Cl).

Pentagonal Hemihedrism. The second kind of possible isometric hemihedrism is, as we have seen (p. 63),

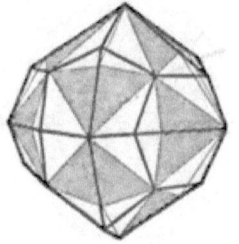
Fig. 74.

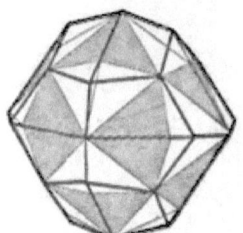
Fig. 75.

produced by the selection of one half of the planes of the hexoctahedron by pairs which intersect in the principal planes of symmetry, or octahedral edges.

The survival and extension of the planes belonging to these two sets of alternating pairs produce from

THE ISOMETRIC SYSTEM.

the most general form two new figures, each bounded by twenty-four similar trapeziums (Figs. 74 and 75). These forms possess three sorts of solid angles and three sorts of edges. The two figures produced by the survival of the two halves of the planes of any hexoctahedron differ from each other only in their position. Either one may, by a revolution of 90° about any of its axes, be brought exactly into the position of the other. Such figures, in contradistinction to enantiomorphous forms, are said to be *congruent*. Their positions are distinguished as positive and negative. The above-described hemihedrons are called *didodecahedrons* or *diploids*. Their symbols according to Naumann and Miller are

$$+\left[\frac{mOn}{2}\right], \quad \pi\{hlk\} \quad \text{and} \quad -\left[\frac{mOn}{2}\right], \quad \pi\{hkl\}.$$

Any other isometric form is capable of producing a geometrically new hemihedron by this method of selec-

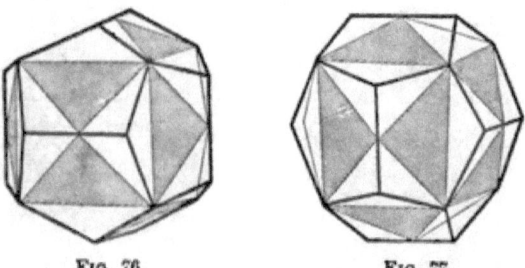

FIG. 76.   FIG. 77.

tion, if its planes correspond exactly to pairs of planes selected on the hexoctahedron. This is the case with the faces of the tetrahexahedron, but with those of no other isometric form. The selection of alternate planes of the tetrahexahedron produces two new and congruent forms, bounded by twelve similar but unequilateral pentagons (Figs. 76 and 77). These are known as the positive and negative *pentagonal dodecahedrons*, or, on account of their frequent occurrence on crystals of pyrite (FeS$_2$), as *pyritohedrons*. Their symbols are written by Naumann and Miller

$$+\left[\frac{\infty On}{2}\right], \quad \pi\{kh0\} \quad \text{and} \quad -\left[\frac{\infty On}{2}\right], \quad \pi\{hk0\}.*$$

Apparently Holohedral Hemihedrons. On no other isometric holohedron, except the two just mentioned, do the planes correspond exactly to the pairs of faces selected on the most general form; therefore, which-

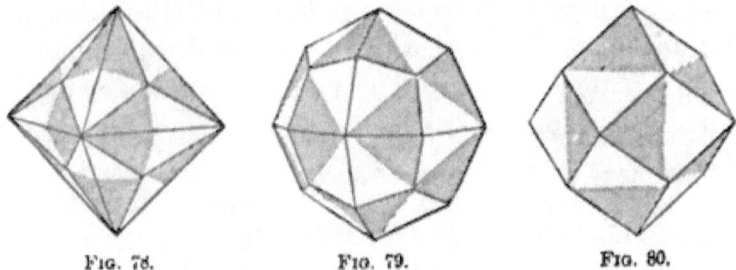

Fig. 78.   Fig. 79.   Fig. 80.

ever pair is selected, portions of all the faces will survive on the five other holohedrons, and these, by their extension, will reproduce the forms as they were before. This is shown by the five figures, 78–82, which are shaded to correspond to the pentagonal

---

* The regular dodecahedron of geometry, with its angles all equal and its faces equilateral pentagons, is not crystallographically possible, because its indices would be $2.1+\sqrt{5}.0$, involving an irrational quantity. The form whose indices are $\pi\{580\}$ approaches it very closely.

selection. As explained in the case of gyroidal hemihedrism, forms of this character are to be regarded

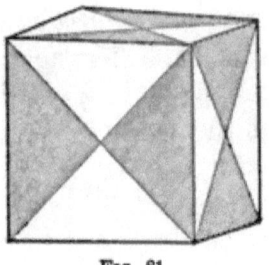

Fig. 81.   Fig. 82.

as truly hemihedral as though they produced completely new shapes.

Symmetry of Pentagonal Forms. An examination of the two new form-types produced by this kind of hemihedrism shows that they possess three planes of symmetry, corresponding in their position to the three principal planes of symmetry of holohedral forms. These are, however, no longer principal planes of symmetry, because they do not contain strictly interchangeable directions (p. 34). The six secondary planes of symmetry belonging to holohedral forms have disappeared.

The pentagonal hemihedrism is also called *parallel-face*, because the planes of its forms, like those of holohedrons, are arranged in parallel pairs. With the succeeding kind of hemihedrism this is, however, not the case, so that this is called *inclined-face*, as has been already explained on p. 42 of the preceding chapter.

Pentagonal Hemihedral Forms in Combination. On account of the common occurrence of parallel-face hemihedrism in the isometric system, it will be well to figure and describe a few of the most frequent combinations. **Fig.** 83 shows the cube with its edges unsymmetrically truncated (p. 38) by the faces of the pentagonal dodecahedron. Fig. 84 shows the same hemihedron blunting the solid angles of the octahedron. When these two forms have about the same development, a combination results which simulates the shape of the regular icosahedron of geometry

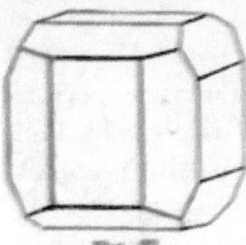

FIG. 83.

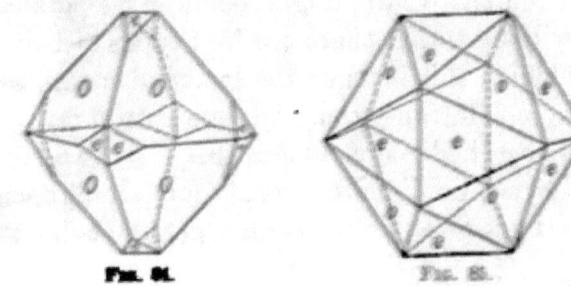

FIG. 84.   FIG. 85.

(Fig. 85). This solid cannot, however, itself be exactly represented by any crystal, because its planes would have irrational indices. Fig. 86 shows a similar

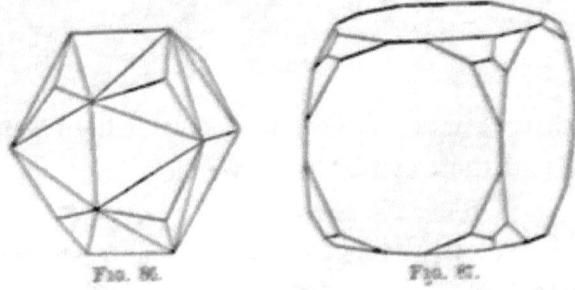

FIG. 86.   FIG. 87.

combination to the last, where the octahedron is replaced by the diploid. Fig. 87 is an example of a cube whose solid angles are replaced by the trihedral angles of the diploid, $\left[\dfrac{3O\tfrac{3}{2}}{2}\right]$, $\pi\{231\}$.

As examples of parallel-face hemihedrism may be mentioned stannic iodide, SnI_4; iron disulphide, FeS_2 (pyrites); cobalt arsenide, $CoAs_2$ (cobaltite); cobalt-glance, $(Co,Fe)AsS$; alum, $\dot{R}\ddot{R}(SO_4)_2 + 12$ aq.

Inclined-face or Tetrahedral Hemihedrism. The selection of planes by alternate octants will produce geometrically new forms from all holohedrons whose faces belong exclusively to single octants. A glance at Plate I will show that there are four types of isometric forms of which this is true: the hexoctahedron, trisoctahedron, icositetrahedron, and octahedron.

The hexoctahedron yields, in this way, two congruent half-forms, bounded by twenty-four similar scalene triangles (Figs. 88 and 89). Both edges and solid angles

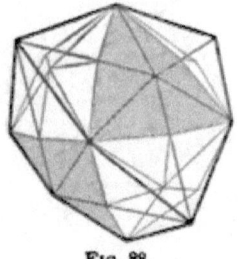 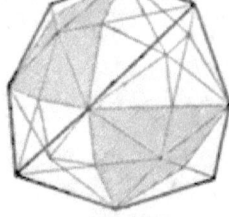

FIG. 88.   FIG. 89.

are of three kinds. These forms are called *hextetrahedrons*, and their symbols are written

$$+\dfrac{mOn}{2},\ \kappa\{hkl\} \quad \text{and} \quad -\dfrac{mOn}{2},\ \kappa\{h\bar{k}l\}.$$

The trisoctahedron, by the same method of selection of its planes, yields two congruent half-forms, bounded by twelve similar trapeziums, intersecting in three kinds of solid angles and two sorts of edges (Figs. 90

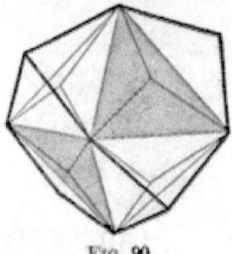
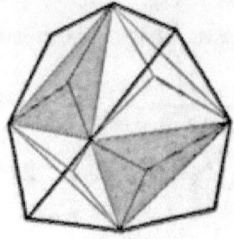

Fig. 90.   Fig. 91.

and 91). They are known as *tetragonal tristetrahedrons*, their symbols being

$$+\frac{mO}{2}, \quad \kappa\{hhl\} \quad \text{and} \quad -\frac{mO}{2}, \quad \kappa\{h\bar{h}l\}.$$

The corresponding half-forms developed from the icositetrahedron are bounded by twelve similar isosceles triangles, intersecting in two kinds of edges and

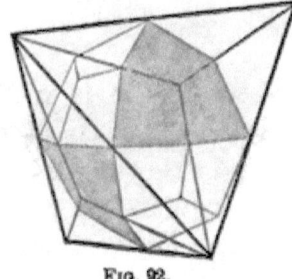
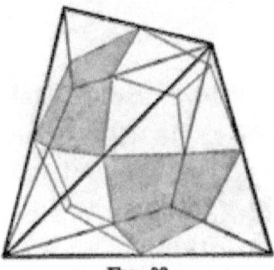

Fig. 92.   Fig. 93.

solid angles (Figs. 92 and 93). These are called *trigonal tristetrahedrons*,* their symbols being

$$+\frac{mOm}{2}, \quad \kappa\{hkk\} \quad \text{and} \quad -\frac{mOm}{2}, \quad \kappa\{h\bar{k}k\}.$$

---

* The trigonal tristetrahedron is also called the *trigondodecahedron* or *pyramid-tetrahedron;* while the tetragonal tristetrahedron is known as the *deltoid dodecahedron*. It is well to call attention to the resem-

The octahedron yields two congruent half-forms, which are the regular *tetrahedrons* of geometry. They are bounded by four equilateral triangles, intersecting in six similar edges and four equal trihedral angles

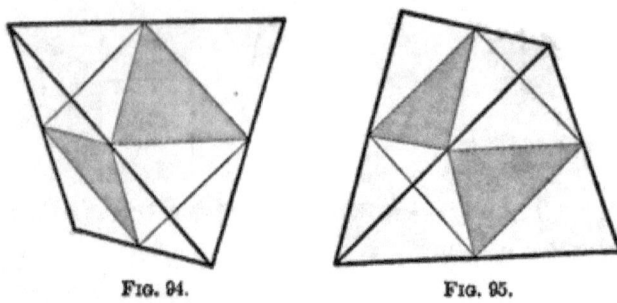

Fig. 94.   Fig. 95.

(Figs. 94 and 95). These are the simplest crystal forms in any system which completely enclose space. Their symbols are

$$+\frac{O}{2}, \ \kappa\{111\} \quad \text{and} \quad -\frac{O}{2}, \ \kappa\{1\bar{1}1\}.$$

Apparently Holohedral Hemihedrons. It is evident that the three other isometric holohedrons (tetrahexahedron, rhombic dodecahedron and cube) whose planes belong equally to two contiguous octants can produce no new forms in the tetrahedral hemihedrism, because the parts of their planes which disappear in one octant are reproduced by the extension of the portions that

---

blance between the planes of the latter hemihedron and those of the holohedron from which the former is derived (icositetrahedron); and *vice versa*. If this cross-resemblance is overlooked, confusion is apt to arise in remembering the forms from which these two hemihedrons are developed.

remain in the next octant, and so on. This will become clear by an inspection of the following figures (96, 97 and 98).

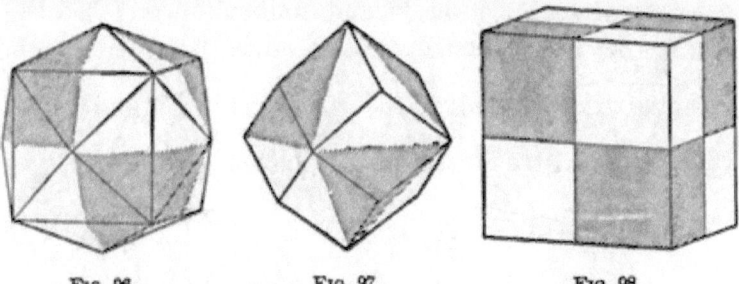

Fig. 96. Fig. 97. Fig. 98.

Symmetry of Tetrahedral Forms. In the geometrically new forms which are produced by the inclined-face or tetrahedral hemihedrism the six secondary planes of symmetry of the isometric system remain, while its three principal planes of symmetry disappear. This may be most easily seen in the case of the tetrahedron (Figs. 94 and 95), all of whose six interfacial angles are bisected by secondary planes of symmetry, corresponding in their positions to the faces of the rhombic dodecahedron.

Tetrahedral Forms in Combination. A few of the more frequent tetrahedral combinations may be men-

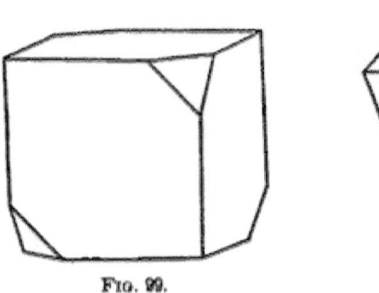
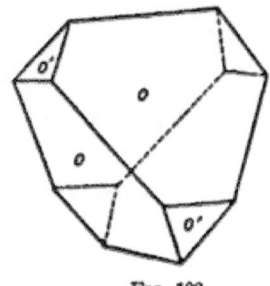

Fig. 99. Fig. 100.

although the cube is not geometrically different from its corresponding holohedron. Fig. 100 shows a positive and negative tetrahedron in combination; Fig. 101, a tetrahedron (*o*), cube (*h*), and dodecahedron (*d*). Fig. 102 shows a tetrahedron (*o*), its edges bevelled by the trigonal tristetrahedron, $+\frac{2O2}{2}$, $\kappa\{211\}$ (*l*), and its angles replaced by the rhombic dodecahedron (*d*). Fig.

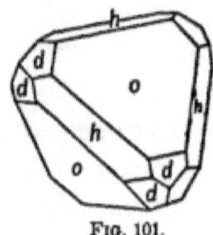
Fig. 101.

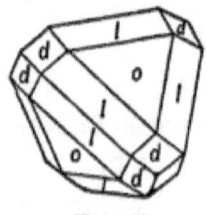
Fig. 102.

103 shows the rhombic dodecahedron (*d*) combined with the trigonal tristetrahedron $+\frac{3O3}{2}$, $\kappa\{311\}$ (*q*), as it sometimes occurs on zinc-blende. Fig. 104 gives another combination, observed on the mineral boracite.

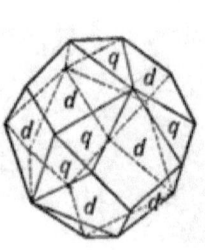
Fig. 103.

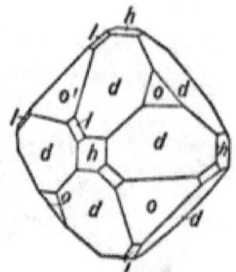
Fig. 104.

It shows the rhombic dodecahedron (*d*), cube (*h*), positive and negative tetrahedrons (*o* and *o'*), and negative trigonal tristetrahedron, $-\frac{2O2}{2}$, $\kappa\{2\bar{1}1\}$ (*l*).

CRYSTALLOGRAPHY.

As prominent examples of tetrahedral crystallization may be mentioned: the diamond; zinc sulphide, ZnS (zinc-blende); sulphide and selenide of mercury, HgS (metacinnabarite) and HgSe (tiemannite); tetrahedrite $\dot{R}_s(AsSb)_2S_7$; magnesium chloroborate, $Mg_7Cl_2B_{16}O_{30}$ (boracite); and the sulpho-silicate, helvine.

Limiting Forms of Isometric Hemihedrons. Hemihedral forms which have a variable parameter oscillate between limiting forms, like the corresponding holohedrons (p. 58). The pentagonal dodecahedron, for instance, approaches the cube in proportion as its parameter, m, is increased; and the rhombic dodecahedron, in proportion as it is diminished. So the trigonal tristetrahedron oscillates between the tetrahedron and the cube; and the tetragonal tristetrahedron, between the tetrahedron and rhombic dodecahedron. All three hemihedral derivatives of the hexoctahedron vary between the limits of all the other isometric forms. These relations are indicated in the three following diagrams:

1. Gyroidal Hemihedrism.   2. Pentagonal Hemihedrism.   3. Tetrahedral Hemihedrism.

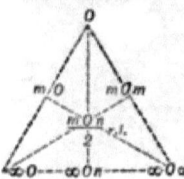 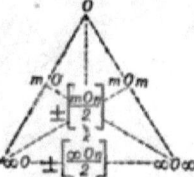 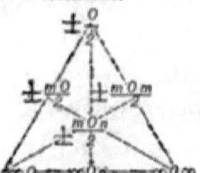

TETARTOHEDRAL DIVISION OF THE ISOMETRIC SYSTEM.

Tetartohedrism, or the independent occurrence of *one quarter* of the planes of a holohedral form (p. 40), may be regarded as due to the simultaneous development of two sorts of hemihedrism. To discover what kinds of tetartohedrism may occur in any crystal system, we may therefore combine its hemihedral selections in every way possible and note which of the results satisfy the required conditions. In the isometric system the result is the same, whichever two hemihedrisms we combine. This will be evident upon an examination of the three adjoining figures (105, 106 and 107) representing the three methods of hemihe-

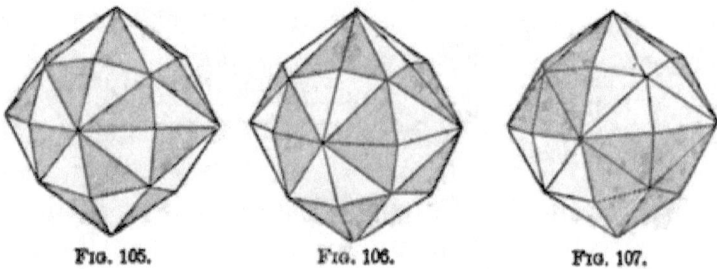

FIG. 105.   FIG. 106.   FIG. 107.

dral selection. Whichever two of these three figures we imagine superposed, the result is the same, viz., three alternate planes surviving in alternate octants, as shown in Fig. 108. If we imagine the twelve white planes on this hexoctahedron to be extended until they mutually intersect, the result will be a new form, bounded by irregular pentagons. There must, of course, be four of these quarter-forms derivable from

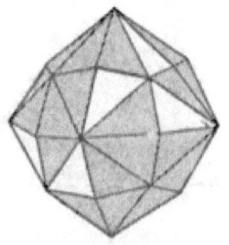

FIG. 108.

every hexoctahedron. The two developed from the alternating planes of the *same* octants will be enantiomorphous (Figs. 109 and 110); and to each of these two forms there will be a congruent form, derived from the sets of planes in the other octants. The enantiomorphous pairs are distinguished as right- and left-handed, and the congruent pairs are distinguished

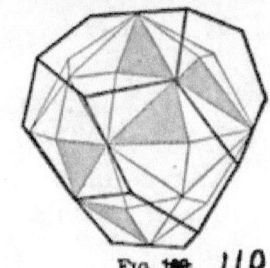

Fig. 110

Fig. 109

as positive and negative. These quarter-forms are called *tetrahedral-pentagonal dodecahedrons*. They are designated as follows:

Positive right-handed, $+\dfrac{m\,O\,n}{4}r$, $\kappa\pi\,\{lkh\}$ ⎫
⎬ congruent pair.
Negative right-handed, $-\dfrac{m\,O\,n}{4}r$, $\kappa\pi\,\{k\bar{l}h\}$ ⎭

Positive left-handed, $+\dfrac{m\,O\,n}{4}l$, $\kappa\pi\,\{klh\}$ ⎫
⎬ congruent pair.
Negative left-handed, $-\dfrac{m\,O\,n}{4}l$, $\kappa\pi\,\{\bar{l}kh\}$ ⎭

The combination of a positive right-handed with a negative left-handed form, or *vice versa*, would produce the *pentagonal icositetrahedron* (gyroidal form). The union of the members of either congruent pair would produce the *diploid* (parallel-face form); while the union of the members of either enantiomorphous pair would produce the *hextetrahedron* (inclined-face form).

No other geometrically new form can result in the isometric tetartohedrism. This will be made clear by

THE ISOMETRIC SYSTEM.

the following ten figures (111-120) which show the result of the simultaneous application of two hemihedrisms

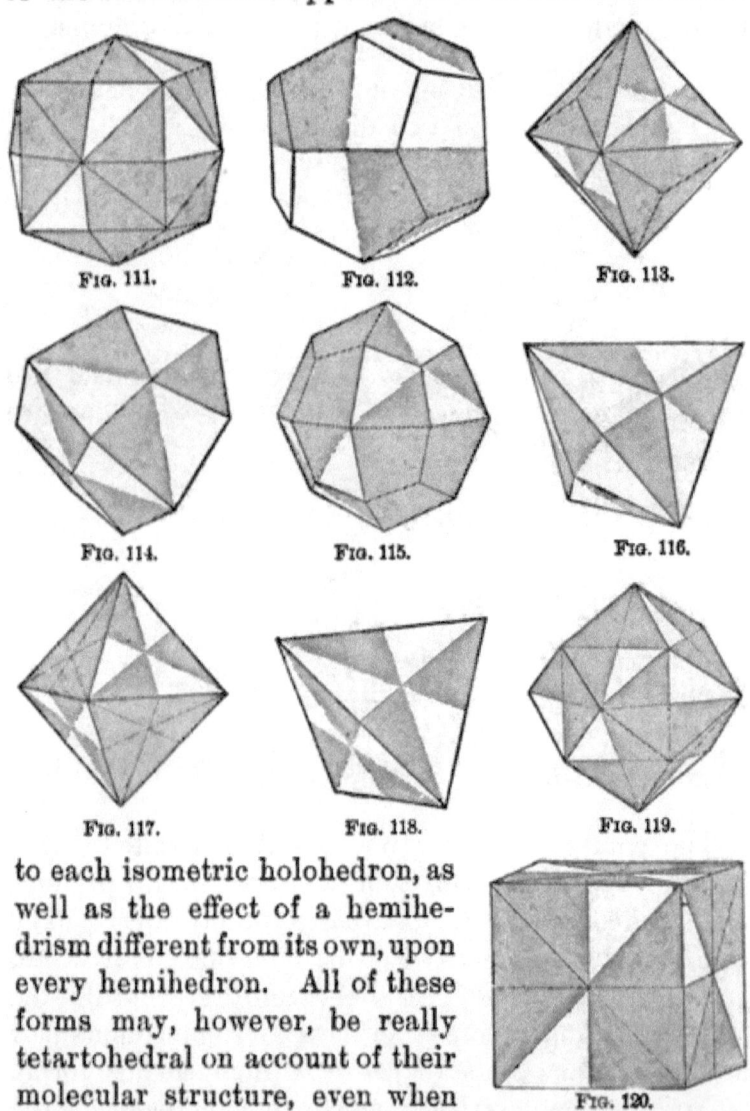

Fig. 111. Fig. 112. Fig. 113.
Fig. 114. Fig. 115. Fig. 116.
Fig. 117. Fig. 118. Fig. 119.
Fig. 120.

to each isometric holohedron, as well as the effect of a hemihedrism different from its own, upon every hemihedron. All of these forms may, however, be really tetartohedral on account of their molecular structure, even when externally they do not differ from hemihedral or holohedral forms.

There are two ways of recognizing a crystal as tetartohedral: (1) by identifying on it the faces of the geometrically tetartohedral form; or (2) by discovering on it faces of forms belonging to two different kinds of hemihedrism. For instance, if we observe on the same crystal the tetrahedron ($-o$) and the pentagonal dodecahedron (p), as may sometimes be done in the case of sodium chlorate (Fig. 121), we may conclude that the substance is tetartohedral, and therefore, that if the most general form occurred on it, it could be represented by but one quarter of its planes.

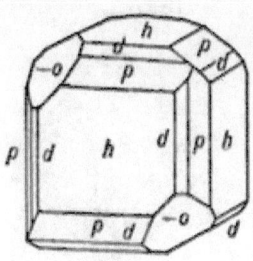

Fig. 121.

Other substances showing isometric tetartohedral crystallization are: sodium bromate; the nitrates of lead, barium, and strontium; and uranyl sodium acetate, $NaUO_2(C_2H_3O_2)_3$. A manifold combination observed on a crystal of barium nitrate is shown in Fig. 122. It has the forms

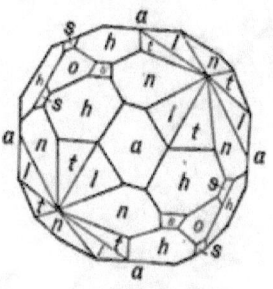

Fig. 122.

$$\infty O \infty, \{100\} (a); \quad -\frac{O}{2}, \kappa\pi \{1\bar{1}1\} (o);$$

$$+\frac{3O3}{2}, \kappa\pi \{311\} (l); \quad -\frac{2O2}{2}, \kappa\pi \{2\bar{1}1\} (s);$$

$$+\frac{4O2}{4}l, \kappa\pi \{214\} (t); \quad -\frac{4O2}{4}r, \kappa\pi \{2\bar{1}4\} (h);$$

and $\quad +\frac{5O\frac{5}{3}}{4}r, \kappa\pi \{351\} (n).$

CHAPTER IV.

THE TETRAGONAL SYSTEM.*

HOLOHEDRAL DIVISION.

Second Class of Crystal Systems. The second class of crystal systems, defined on p. 44, comprises all the forms which possess a single principal plane of symmetry. It is customary to place this principal plane, to which all of the secondary planes of symmetry are perpendicular, in a horizontal position. In a physical sense, the symmetry of the two systems belonging to this class is the same, and the physical properties of all of their crystals are identical. Their differences of form are such as arise from the presence of *two* lateral axes intersecting at 90°, or of *three* lateral axes intersecting at 60°.

On account of its higher grade of symmetry it might seem more logical to consider the hexagonal system first. It is nevertheless deemed advisable to give the tetragonal system precedence, inasmuch as it is the easier of the two to understand; and, at the same time, the more closely related to the isometric system.

Symmetry. The distribution of the planes of sym-

---

* Also called the *quadratic*, *pyramidal*, or *quaternary* system.

metry belonging to complete or holohedral tetragonal crystals is represented by Fig. 123. There is one principal plane of symmetry (placed horizontally) which is intersected by four vertical secondary planes of symmetry. These latter meet at angles of 45°, in a common line—the principal axis of symmetry. Alternate secondary planes of symmetry only are crystallographically interchangeable, while contiguous planes are not interchangeable. The interchangeable planes must therefore intersect at angles of 90°; and either pair of these may be regarded as determining, by their intersections with the principal plane of symmetry, the directions of the two lateral axes. Whichever two of the secondary planes of symmetry are so regarded are called the *axial planes*, while the other two are then known as the *intermediate planes*.

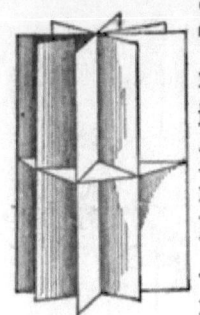

Fig. 123.

Axes. The directions of the tetragonal axes of reference are fully determined by the symmetry of the system. There must be one *principal axis* (German, *Hauptaxe*), normal to the principal plane of symmetry, and therefore vertical in its position. There must be two secondary or *lateral axes* (German, *Nebenaxen*), normal to the principal axis and to each other, whose position is determined by whichever pair of alternate planes of symmetry have been selected as axial planes. The intersections of the other two secondary planes of symmetry with the principal plane determine two directions which are sometimes referred to as the *intermediate axes* (German, *Zwischenaxen*).

The directions of the tetragonal axes are therefore

the same as those of the isometric system; their essential distinction from the latter is, however, that while the lateral axes are still equal and interchangeable with each other, they are no longer interchangeable with the principal or vertical axis. We must therefore designate the vertical axis by a different letter, c, from that used for the lateral axes, although the use of signs is the same as has been explained for the isometric system (Fig. 124).

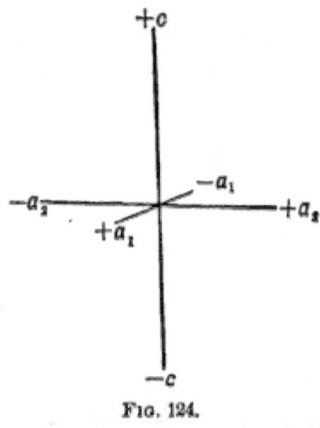

Fig. 124.

The Fundamental Form and Axial Ratio. The fundamental or ground-form (German, *Grundform*) has been defined (p. 47) as composed of those planes which intercept all the axes at their unit lengths. In the isometric system, the equality of all the axes precludes the possibility of more than one ground-form; when, however, as in the present case, the lateral and vertical axes are irrational multiples of each other, there may be a variety of fundamental forms in the same system, just as there may be a variety of irrational inequalities between the axes. The ratio existing between the unit lengths of the two unequal axes a and c is expressed by the quotient $\frac{c}{a}$, where a is assumed as unity. This quotient is called the *axial ratio*, and is a very important crystallographic quantity. The axial ratios derived from all forms occurring on crystals of the same substance under the same conditions must, according to the law of rationality of the indices (p. 26), be rational multiples of one another, while those derived from forms on crystals of different substances are irrational multiples of one another. Thus the axial ratio becomes a physical constant for all crystallized matter that is not isometric, and serves to identify it in the same way that specific gravity, hardness, or elasticity does.

It is important to determine accurately the axial ratio for different substances. This is easily accomplished for tetragonal crystals if we know by measurement the values of either of the interfacial angles belonging to the ground-form. Thus (Fig. 125) if we assume the length of the lateral axis a as equal to unity, the length of c, expressed in terms of a, is $=$ tg b; while $\sin b = \cotg A = \cotg \frac{1}{2} X$, or tg $b = $ tg $\frac{1}{2} Z \sqrt{\frac{1}{2}}$.

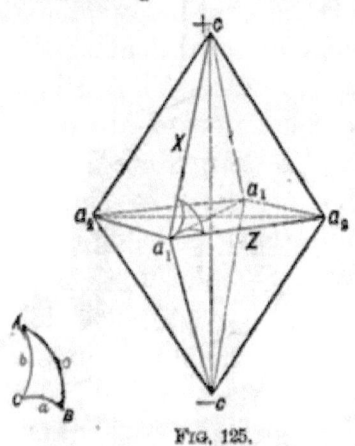

Fig. 125.

Mere inequality in the length of the intercepts on the lateral and vertical axes is not alone enough to constitute a tetragonal form. The ratio existing between these unequal intercepts must further be an irrational quantity. A plane, for instance, whose symbol is $a : a : 2a$ has an intercept on one axis which is unequal to those upon the other two axes, but this is the plane of an isometric form so long as the parameter, 2, is a rational multiple of *all* the axes. If, however, the parameter, 2, is only a rational multiple of the vertical axis, but an irrational multiple of the lateral axes, the symbol must be written $a : a : 2c$, and the plane becomes truly tetragonal. Hence the axial ratio is an irrational quantity which is characteristic of a given chemical compound.

When two or more tetragonal pyramids occur on crystals of the same substance, it is necessary to select one of them as the ground-form, and from this to calculate the axial ratio. It is not a matter of great importance which particular pyramid is selected for this purpose, since, according to the law of rational indices, the intercepts of all planes on the same or equivalent axes must be even multiples of one another. The ratio between the intercepts of *any* plane on dissimilar axes must therefore always be the axial ratio, or some even multiple of it, for the particular substance to which the plane belongs. To secure uniformity in the choice of a ground-form, it is, however, customary to select as such the most common or most prominently developed pyramid, or else one that is distinguished by some physical property like cleavage.

Development of the possible Holohedral Forms in the Tetragonal System. The inequality in the unit lengths of the vertical and lateral tetragonal axes signifies that these axes are crystallographically dissimilar and not interchangeable—a fact which is also directly derivable from the symmetry of the system (p. 82). The most general parameter symbol for a tetragonal form is therefore

$$na_1 : a_2 : mc,$$

which is evidently capable of only two permutations, if only a_1 and a_2 are interchangeable:

$$na_1 : a_2 : mc \quad \text{and} \quad a_1 : na_2 : mc.$$

Hence the most general tetragonal symbol stands for two planes in each octant, or sixteen planes in all.

The most general index symbol (hkl) leads us to the same result, when only h and k are interchangeable. With every possible combination of signs, representing the different octants, we obtain

Four upper octants.				Four lower octants.			
hkl	$\bar{h}kl$	$\bar{h}\bar{k}l$	$h\bar{k}l$	$hk\bar{l}$	$\bar{h}k\bar{l}$	$\bar{h}\bar{k}\bar{l}$	$h\bar{k}\bar{l}$
khl	$\bar{k}hl$	$\bar{k}\bar{h}l$	$k\bar{h}l$	$kh\bar{l}$	$\bar{k}h\bar{l}$	$\bar{k}\bar{h}\bar{l}$	$k\bar{h}\bar{l}$

The same number of planes for the most general tetragonal form is also indicated by the symmetry of the system. From Fig. 123 we see that the five planes of symmetry divide space into sixteen similar secants, so that any crystal face, oblique to all of these planes, necessitates another in each of the other fifteen secants.

The other possible tetragonal holohedrons may be derived from the most general formula $na_1 : a_2 : mc$, by assigning limiting values to the two parameters, n and m. Inasmuch as it is customary to make the *lesser* lateral intercept equal to unity, the limiting values for n are one and infinity. Since m, however, refers only to a single axis, its value may vary between zero and infinity.

By assigning these limiting values first to *one* and then to *both* of the parameters, we obtain seven and only seven possbile tetragonal form-types, which, as in the isometric system (p. 50), fall naturally into three classes, as follows:

Class I. Forms with *two* variable parameters.

   1. $m \gtreqless n$, general symbol becomes $na_1 : a_2 : mc$.

Class II. Forms with *one* variable parameter.

   2. $n = 1$, general symbol becomes $a_1 : a_2 : mc$.

3. $n = \infty$, general symbol becomes $\infty a_1 : a_2 : mc$.
4. $m = \infty$, general symbol becomes $na_1 : a_2 : \infty c$.

Class III. Forms with *no* variable parameter.

5. $m = \infty$ and $n = 1$, general symbol becomes
$$a_1 : a_2 : \infty c.$$
6. $m = \infty$ and $n = \infty$, general symbol becomes
$$\infty a_1 : a_2 : \infty c.$$
7. $m = 0$ and $n = 1$, general symbol becomes
$$a_1 : a_2 : 0c.$$

Although identical in their number, these tetragonal form-types do not correspond exactly with those of the isometric system in character, as we shall see from a more special description of the tetragonal holohedrons.

The Ditetragonal Pyramid. The most general tetragonal symbol, $na_1 : a_2 : mc$, stands for a double pyramid containing two planes in each octant (Fig. 126). The vertical intercepts for all the planes of this form must be the same, because the vertical axis is not interchangeable with the others. This form is called the *ditetragonal pyramid*. Like the most general isometric form, it has three kinds of edges and three sorts of solid angles. The edges which lie in the principal plane of symmetry are called *basal edges* and are all similar.

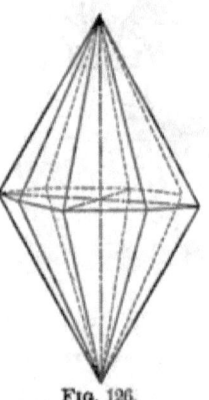

Fig. 126.

Those which connect the lateral and vertical axes are called *polar edges*, and are alternately dissimilar. The faces of the ditetragonal pyramid are all similar scalene triangles. Its symbol according to the notation of Naumann is mPn; its general index symbol is $\{hkl\}$.

Tetragonal Pyramids of the First and Second Order. If, in the parameter symbol of the most general tetragonal form, $na_1 : a_2 : mc$, the parameter n be given its two limiting values, while the parameter m is allowed to remain unchanged, two new eight-sided pyramids result which are the limiting forms of the ditetragonal pyramid in two directions.

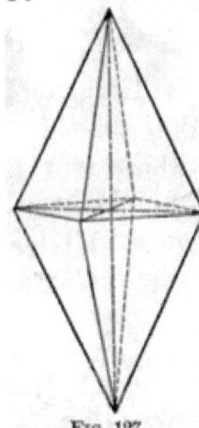

Fig. 127.

If n be made equal to unity, the formula becomes $a_1 : a_2 : mc$, which represents a form each of whose planes occupies a single octant (Fig. 127). This is called the *tetragonal pyramid of the first order*. Its faces are similar isosceles triangles, and its edges and solid angles are both of two kinds. Its symbols, according to Naumann and Miller, are mP and $\{hhl\}$.

If, on the other hand, n be given its maximum value, infinity, the formula becomes $\infty a_1 : a_2 : mc$, which represents another tetragonal pyramid, each of whose planes is common to two octants (Fig. 128). This is called the *tetragonal pyramid of the second order*. Its symbols, according to Naumann and Miller, are $mP\infty$ and $\{0hl\}$.

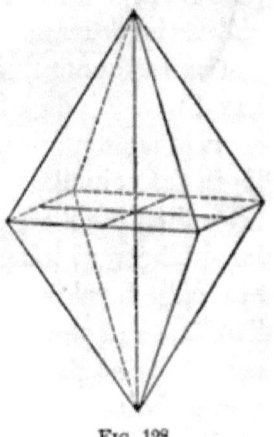

Fig. 128.

The diagram, Fig. 129, illustrates the relations of the three tetragonal pyramids, as shown in their cross-sections. The inner square gives the position of the pyramid of the first order with reference to the axes, and the outer square that of the pyramid of the second order. The octagon between these two squares gives the intermediate position of the ditetragonal pyramid. The pyamids of the first and second orders differ only in their relative positions. The order of either depends entirely on which of the two pairs of interchangeable axes (p. 82) we select as lateral axes.

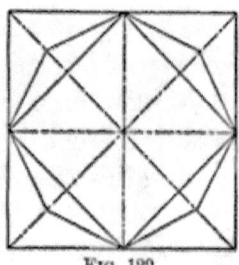
Fig. 129.

No new type of tetragonal form is obtained by giving the vertical parameter, m, the value unity, because this is not a limiting value for the vertical axis. It is, however, customary to designate those pyramids whose vertical parameters are unity as *unit pyramids*. They do not in reality differ from the other pyramids, any more than these differ among themselves, since it is in a measure arbitrary which of the pyramids belonging to a substance is selected as the ground-form. The parameter symbols of these unit pyramids are written Pn, P, and $P\infty$; and their indices $\{hkk\}$, $\{111\}$ and $\{101\}$.

Tetragonal Prisms. If we give the lateral parameters of the tetragonal formula the values they possess in

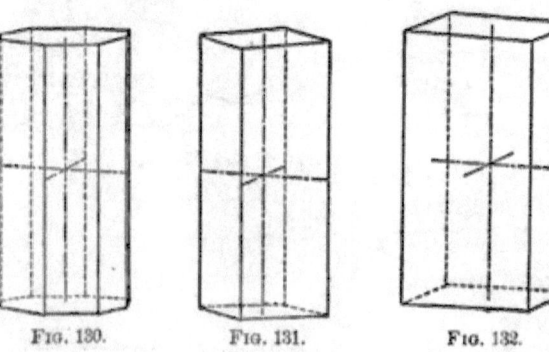
Fig. 130. Fig. 131. Fig. 132.

crease the vertical parameter, m, to its maximum limit, infinity, three types of prisms will result, which correspond in all respects to the three pyramids, except that their planes are parallel to the vertical axis (Figs. 130, 131, and 132). These are all open forms (p. 36) and they cannot therefore occur except in combination. Their relative positions are shown in Fig. 129. They are named, to accord with the corresponding pyramids: the *ditetragonal prism*, ∞Pn, $\{hk0\}$; the *tetragonal prism of the first order*, ∞P, $\{110\}$; and the *tetragonal prism of the second order*, $\infty P \infty$, $\{100\}$.

The last-named form, in spite of its being called a prism, really belongs to the type of pinacoids (p. 36); just as the pyramid of the second order really belongs to the type of prisms, since its planes are parallel to one of the axes.

The Basal Pinacoid. The last of the seven possible types of tetragonal forms is produced by giving the vertical parameter, m, its minimum limiting value, zero. This causes all of the pyramidal types to merge into one plane whose position is that of the principal plane of symmetry (Fig. 123). When such a plane, whose parameter symbol is $a_1 : a_2 : 0c$, is divided into the pair of parallel planes necessary to produce a holohedral form (p. 18), they have the position shown in Fig. 133, parallel to both lateral axes. These planes are of unlimited extent, and can hence only occur in combination with other forms. They are called *basal pina-*

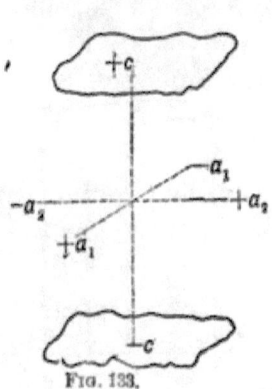

Fig. 133.

coids, and are designated in the notations of Naumann and Miller by the symbols $\frac{1}{2}OP\{001\}$ (p. 29).

Limiting Forms in the Tetragonal System. The relation of limiting forms among the tetragonal holohedrons may be illustrated by the following diagram, where m indicates a vertical parameter greater than unity, and $\frac{1}{m}$ one less than unity.

OP	OP	OP	Pinacoid.
$\frac{1}{m}P$ ---	$\frac{1}{m}Pn$ ---	$\frac{1}{m}P\infty$	
P ---	Pn ---	$P\infty$	Pyramids.
mP ---	mPn ---	$mP\infty$	
∞P ---	∞Pn ---	$\infty P\infty$	Prisms.

Limiting forms within a single system are produced by variations in *parameters*, until these reach a fixed or limiting value. This we have seen illustrated in both the isometric and tetragonal systems, and we shall find that it is equally true of all the other systems. There is, however, another way in which limiting forms may be produced, and that is by variations of the *axial ratio*. In such cases the limiting forms always belong to another system, of a higher grade of symmetry than that possessed by the forms which they limit. For instance, the first of the two following figures (134) represents a tetragonal pyramid, which remains such as long as the vertical axis is either greater or less than the lateral axes, no matter how small the difference may be. In the case of the iron-copper-sulphide, chalcopyrite, the ratio $a : c$ is $1 : 0.9856 +$, which is very near $1 : 1$. If this limit were actually reached, however, the form would cease to be tetragonal, and would become its limiting form in the isometric system—the octahedron (Fig. 135), whose interfacial angles are all 109° 28′ 16″.4, instead of 108° 40′ and

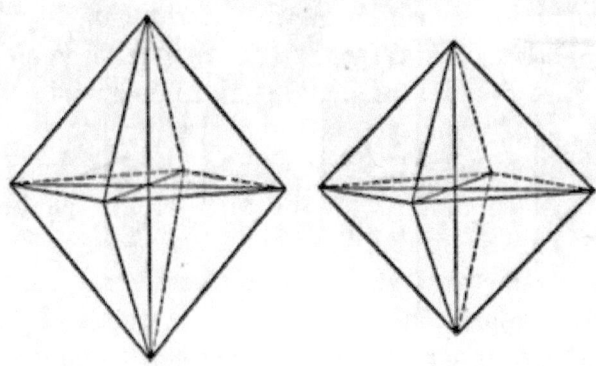

Fig. 134.   Fig. 135.

109° 53′ as they are in chalcopyrite. Many crystals approach so closely to their limiting forms in systems of a higher grade of symmetry that their true character can only be discovered by an examination of their optical properties, or by some other physical test more delicate than the measurement of their interfacial angles. Such a close approach, on the part of any crystal, to a grade of symmetry higher than it really possesses is called by Tschermak *pseudo-symmetry*.

The Crystal Series. The whole sequence of possible tetragonal holohedral forms, as above developed, may evidently belong to a single set of axes. In other words, their nature and existence is dependent upon the symmetry of the system, but is not dependent upon the axial ratio. But such planes only can occur on crystals of a given substance as satisfy the law of rational indices for its particular axial ratio. Inasmuch, however, as there are as many distinct axial ratios in the tetragonal system as there are chemical substances crystallizing with the tetragonal symmetry (p. 82), it is plain that each substance has its own tetragonal forms which represent the same types as those of other substances and yet which differ from them in their exact axial inclinations and interfacial angles.

Such a difference necessitates the introduction of a new group of crystal forms called the *Crystal Series*. This may be defined as the sum of all the crystal forms which are possible upon the same set of axes or with the same axial ratio. Each series may contain a complete representation of all the types of forms, while the number of different series is equal to the number of different substances crystallizing in the system.

In the isometric system it is evident that distinct crystal series are impossible, since for all substances the axial ratio is the same.

Holohedral Tetragonal Forms in Combination. The simpler tetragonal combinations are readily intelli-

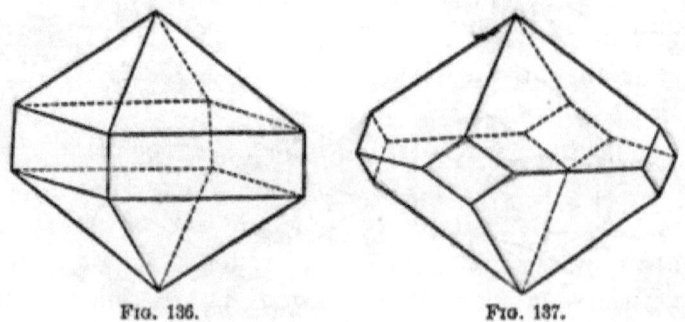

FIG. 136.   FIG. 137.

gible, as will be seen from the accompanying illustrations. Fig. 136 shows the union of a pyramid and prism of the same order, while Fig. 137 gives the result of a union of the same forms belonging to different orders. The next three figures show combinations of pyramids of the first and second orders—Fig. 138

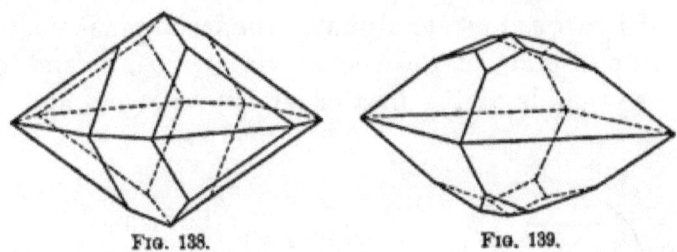

FIG. 138.   FIG. 139.

when the two have equal vertical parameters; Fig. 139 when the pyramid of the second order is the more obtuse, and Fig. 140 when it is the more acute of the two forms.

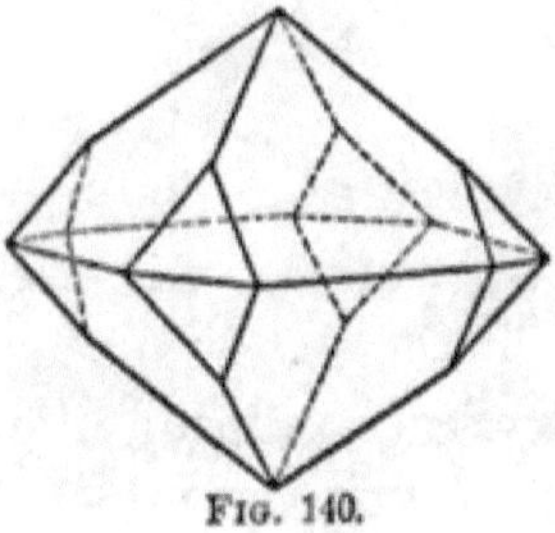

The four succeeding figures represent certain more complex tetragonal combinations. Fig. 141 shows a crystal of manganese dioxide, MnO_2 (polianite), bounded by the unit pyramid of the second order,

FIG. 140.

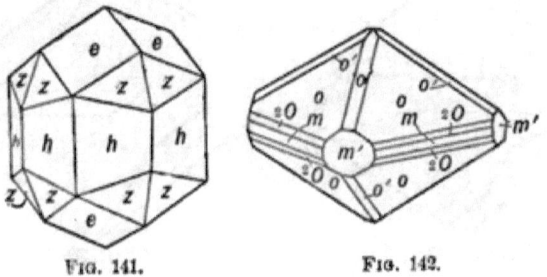

FIG. 141.   FIG. 142.

$P\infty$, $\{101\}$ (e), the ditetragonal pyramid $3P\frac{3}{2}$, $\{321\}$ (z), and the ditetragonal prism $\infty P2$, $\{210\}$ (h). Fig. 142 represents a crystal of boron with the unit pyramids of both orders, P, $\{111\}$ (o) and $P\infty$, $\{101\}$ (o'), the steeper pyramid of the first order, $2P$, $\{221\}$ (2o), and the prisms of both orders ∞P, $\{110\}$ (m) and $\infty P\infty$, $\{100\}$(m'). Fig. 143 gives the planes observed on a crystal of hydrous nickel sulphate: the basal pinacoid, (c); three pyramids of the second order, (m), (o) and (q); two pyramids of the first order, (n) and (p); and the

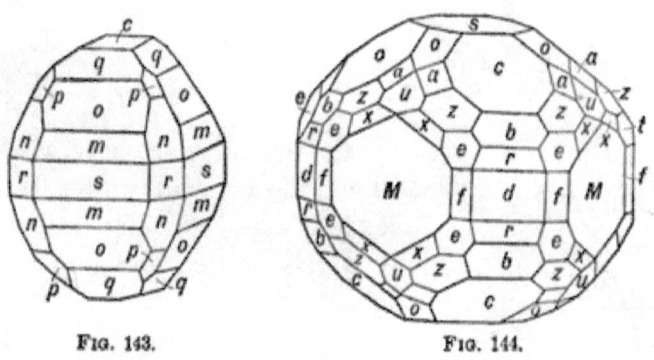

FIG. 143.   FIG. 144.

prisms of both first and second orders, (r) and (s). Finally, in Fig. 144 we have a very complex combination of thirteen tetragonal forms occurring on the mineral vesuvianite: the prism of the first order, (d); prism of the second order, (M); basal pinacoid, (s); three pyramids of the first order, P, $\{111\}$ (c), $2P$, $\{221\}$ (b), and $4P$, $\{441\}$ (r); two pyramids of the second order, $P\infty$, $\{101\}$ (o) and $2P\infty$, $\{201\}$ (u); the ditetragonal prism, $\infty P2$, $\{210\}$ (f); and the four ditetragonal pyramids, $2P2$, $\{121\}$ (z), $4P4$, $\{141\}$ (x), $4P2$, $\{241\}$ (e), and $\frac{3}{2}P3$, $\{132\}$ (a).

As other examples of holohedral tetragonal crystallization may be mentioned stannic oxide, SnO₂ (cassiterite); titanium dioxide, TiO₂, in two of its forms, rutile and anatase; zircon, ZrSiO₄; the chloride, iodide, and cyanide of mercury; and the hydrous silicate, apophyllite.

HEMIHEDRAL DIVISION OF THE TETRAGONAL SYSTEM.

Possible Kinds of Tetragonal Hemihedrism. A careful inspection of the most general tetragonal form—the ditetragonal pyramid—shows that it is possible to select one half of its planes in three different ways so as to satisfy the conditions of hemihedrism given on p. 41. These three methods of selection give rise to three distinct kinds of hemihedrism, which are closely

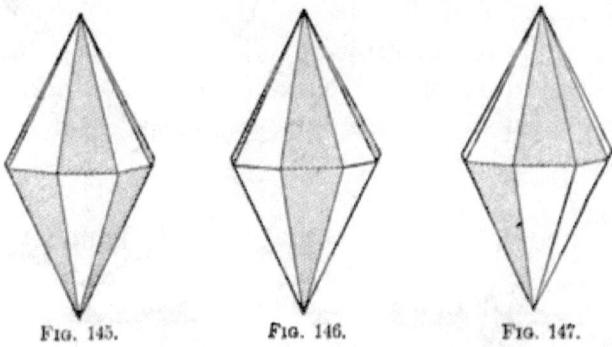

FIG. 145.   FIG. 146.   FIG. 147.

analogous to the three kinds of isometric hemihedrism. The possible methods of selection are represented in the three annexed figures (145, 146 and 147). They are (1) by alternate planes, (2) by pairs of planes intersecting in the principal plane of symmetry, and (3) by alternate octants.

The first produces forms devoid of all symmetry and therefore enantiomorphous. It is called the *trapezohedral hemihedrism.* The second produces forms having one principal, but no secondary planes of symmetry, and is called the *parallel-face* or *pyramidal hemihedrism.* The third produces forms having two secondary, but no principal planes of symmetry, and is called the *inclined-face* or *sphenoidal hemihedrism.*

Trapezohedral Hemihedrism. The extension of alternate planes of the ditetragonal pyramid until they intersect produces an asymmetrical figure bounded by eight similar trapeziums, which meet in three kinds of edges and two kinds of solid angles. Two similar but enantiomorphous forms of this kind are derivable from every ditetragonal pyramid (Figs. 148 and 149). They are called *tetragonal trapezohedrons.* Their general symbols, according to Naumann and Miller, are

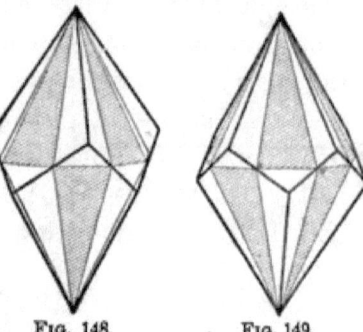

FIG. 148.   FIG. 149.

$$\frac{mPn}{2}r, \quad \tau\{khl\} \quad \text{and} \quad \frac{mPn}{2}l, \quad \tau\{hkl\}.$$

It is evident that no other new form can result by this method of selection from any of the more special tetragonal forms, because, in each of them, the planes correspond to at least two contiguous planes of the most general form.

This mode of crystallization has never been observed on natural minerals, but it has several representatives among organic salts. Examples of these are sulphate of strychnine, sulphate of ethylendiamine, carbonate of guanidine, and diacetylphenol phtalline.

Pyramidal Hemihedrism. The extension of alternate planes on the upper half of the ditetragonal pyramid, and of those directly below them on the lower half, as indicated in Fig. 146, p. 96, must produce a tetragonal pyramid, which is in all respects like that of the first or second order, except in its position. While the two holohedral tetragonal pyramids differ 45° in their positions, this hemihedral tetragonal pyramid, called the *pyramid of the third order*, occupies an intermediate place, which is determined by the exact parameters of the ditetragonal

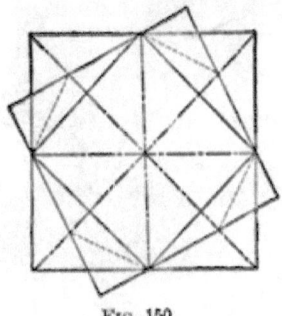

Fig. 150.

pyramid from which it is derived. This may be best illustrated by the adjoining diagram (Fig. 150), which shows the relative positions of the tetragonal pyramids in cross-section, like Fig. 129, except that the alternate faces of the ditetragonal form have here been extended to intersection. The new square thus formed represents the cross-section of the pyramid of the third order. The symbols of the corresponding half-forms derivable from any ditetragonal pyramid are

$$+\left[\frac{mPn}{2}\right], \quad \pi\{khl\} \quad \text{and} \quad -\left[\frac{mPn}{2}\right], \quad \pi\{hkl\}.$$

The same modification is produced by this method of hemihedral selection upon the ditetragonal prism, since each of its planes corresponds to the pair of planes selected on the most general form. Thus two hemihedral *prisms of the third order* result from each ditetragonal prism, whose intermediate positions are likewise shown by Fig. 150. Their symbols are

$$+\left[\frac{\infty Pn}{2}\right], \quad \pi\{kh0\} \quad \text{and} \quad -\left[\frac{\infty Pn}{2}\right], \quad \pi\{hk0\}.$$

If we compare the other tetragonal holohedrons with the most general form, we shall readily see that each of their planes corresponds to two contiguous planes in the upper half of the crystal, and hence that both pyramids and prisms of the first and second order must reproduce themselves in the pyramidal hemihedrism without geometrical alteration. Two new forms only are possible by parallel-face hemihedrism in the tetragonal system, as was also found to be the case in the isometric system.

The tetragonal forms of the third order manifest their true character in combination with other forms, as may be seen in the following examples. Fig. 151 shows a crystal of lead tungstate, $PbWO_4$, (stolzite), bounded by the unit pyramid of the first order, P, $\{111\}$ (o), and by the prism of the third order, $+\left[\frac{\infty P\frac{4}{3}}{2}\right], \pi\{430\}$ (p). Fig. 152 represents a crystal of yttrium niobate (fergusonite) with the basal pinacoid, OP, $\{001\}$ (c); the unit pyramid, P, $\{111\}$ (s); and the pyramid and prism of the third order,

$$-\left[\frac{3P\frac{3}{2}}{2}\right], \pi\{231\} \ (z) \quad \text{and} \quad -\left[\frac{\infty P\frac{3}{2}}{2}\right], \pi\{230\} \ (r).$$

Fig. 153 shows a combination of forms sometimes observed on the silicate scapolite, with the unit pyramid, P, $\{111\}$ (o); the prisms of the first and second orders, ∞P, {110} (M) and $\infty P\infty$, {100} (b); and the pyramid of the third order, $+\left[\dfrac{3P3}{2}\right]$, π {311} (s).

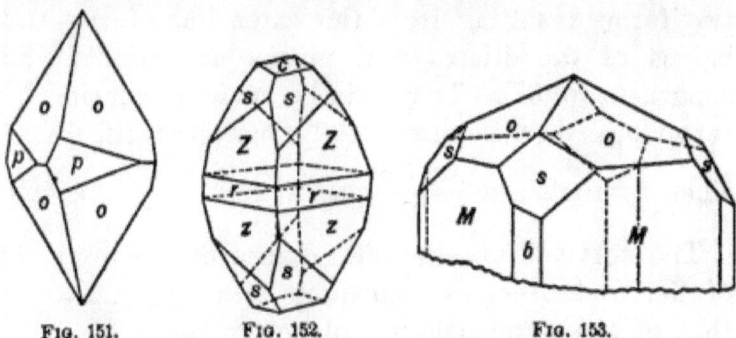

Fig. 151.   Fig. 152.   Fig. 153.

Other examples of tetragonal substances showing pyramidal hemihedrism are calcium tungstate, $CaWO_4$ (scheelite); lead molybdate, $PbMoO_4$ (wulfenite); hydrous magnesium borate, $MgB_2O_4 + 3aq$ (pinnoite); erythroglucine ($C_4H_{10}O_4$) and toluosulphanide.

Sphenoidal Hemihedrism. The selection of planes by alternate octants, as indicated by Fig. 147 on p. 96 for the sphenoidal hemihedrism, can produce only two geometrically new forms in the tetragonal system, since on only two holohedrons of this system do the planes belong exclusively to a single octant. These two forms are the ditetragonal pyramid and the pyramid of the first order.

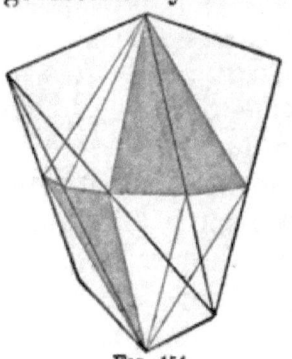

Fig. 154.

The result of extending the pairs of planes occupying alternate octants on the most general form until they intersect is shown in Fig. 154. The figure thus produced is

THE TETRAGONAL SYSTEM. 101

bounded by eight similar scalene triangles, intersecting in three sorts of edges and two sorts of solid angles. It is called the *tetragonal scalenohedron*. The two forms resulting from the extension of the two halves of the ditetragonal planes are similar and congruent (p. 67). They may be made to coincide by revolving either one through 90° about its vertical axis. Their symbols are $+\frac{mPn}{2}$, $\kappa\{hkl\}$ and $-\frac{mPn}{2}$, $\kappa\{h\bar{k}l\}$.

The survival of alternate planes on the pyramid of the first order is analogous with that of the corresponding planes on the isometric octahedron to produce the tetrahedron (p. 73). The result is a form bounded by four isosceles triangles, intersecting in two sorts of edges and with one kind of solid angles (Fig. 155). This is called the *tetragonal sphenoid*, and the symbols of the two congruent half-forms are

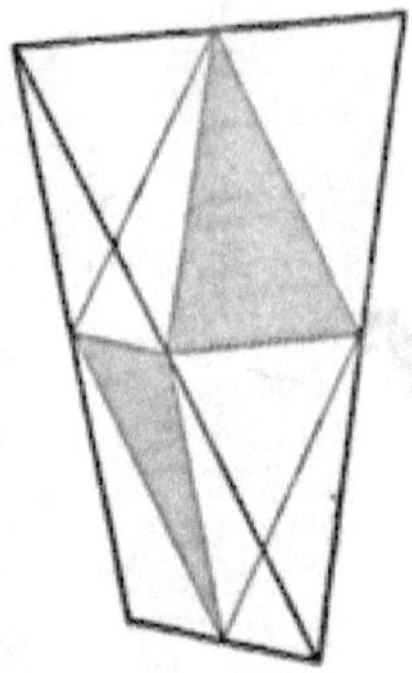

FIG. 155.

$$+\frac{mP}{2}, \quad \kappa\{hhl\} \quad \text{and} \quad -\frac{mP}{2}, \quad \kappa\{h\bar{h}l\}.$$

The sphenoid is well calculated to exhibit the inclined-face character of this hemihedrism. There are acute and obtuse sphenoids, according as the vertical axis is longer or shorter than the lateral axes, and between these two classes the isometric tetrahedron stands as a limiting form of each.

These two new sphenoidal hemihedrons have lost the three axial planes of symmetry belonging to the tetragonal system (Fig. 123), but they retain the two intermediate planes of symmetry.

All other tetragonal forms must appear on crystals of sphenoidal substances as apparent holohedrons, because they are incapable of geometrical modification by this method of selection.

As examples of sphenoidal crystallization may be cited the iron-copper-sulphide, FeS_2Cu (chalcopyrite)

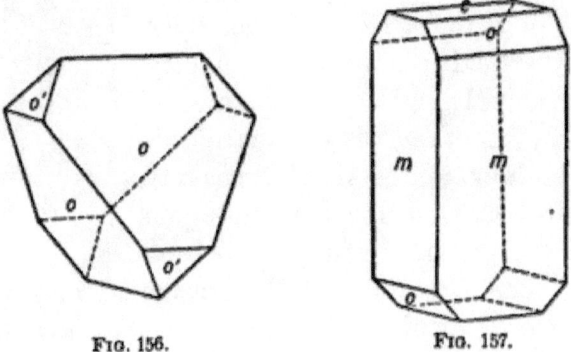

FIG. 156.   FIG. 157.

$\left(\text{Fig. 156, } \pm \frac{P}{2}\right)$, and urea (CH_4N_2O) $\left(\text{Fig. 157, } OP (c),\right.$ $\left.\infty P (m), \text{ and } +\frac{P}{2} (o).\right)$

Tetartohedrism in the Tetragonal System. The simultaneous occurrence of two different kinds of tetragonal hemihedrism is theoretically capable of producing tetartohedral forms, which may at any time be found on natural crystals. A consideration of Figs. 145, 146, and 147 (p. 96) will show that the superposition of the last upon either of the other two will produce a survival of one quarter of the holohedral planes in such a manner as to satisfy the conditions of tetartohedrism (p. 41); while the superposition of the second of these figures upon the first will produce hemimorphism in the direction of the vertical axis.

system only in having six vertical secondary planes, meeting at angles of 30° (Fig. 159), instead of four secondary planes meeting at angles of 45° (Fig. 123, p. 82). Here also only alternate secondary planes are crystallographically equivalent and interchangeable, so that we have again two sets of three *axial* and three *intermediate* vertical planes of symmetry. The principal and axial planes of symmetry together divide space into twelve similar wedge-shaped secants, called *dodecants*, which are analogous to the isometric and tetragonal octants.

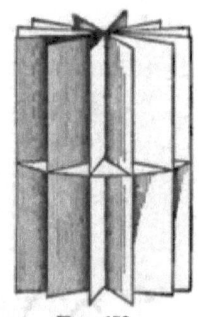

FIG. 159.

Axes. The directions of the hexagonal axes of reference are determined by the planes of symmetry, just as they are in the tetragonal system. The principal axis of symmetry (c) is employed as the vertical axis, while the intersections of either of the two sets of secondary planes of symmetry with the principal plane give three equal lateral axes, all normal to the principal axis and inclined 60° and 120° to one another. It is customary to designate the equal lateral axes by the letters a_1, a_2 and a_3, in the order indicated in Fig. 160; and to assign to their extremities, alternately, plus and

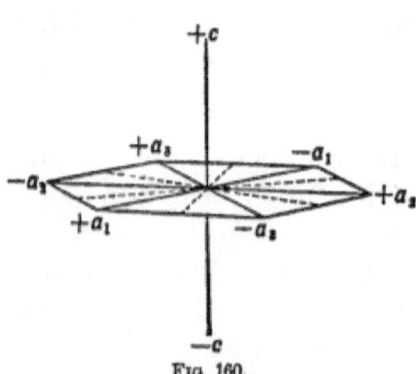

FIG. 160.

minus signs, as first suggested by Bravais.* The intersections of the intermediate planes of symmetry with the principal plane give three additional directions which bisect the angles between the lateral axes and may be called the *intermediate axes*. Their position is shown by the dotted lines in Fig. 160.

The Fundamental Form and Axial Ratio. The significance of these terms has already been fully explained in speaking of the tetragonal system. The hexagonal fundamental or ground-form is bounded by planes which intersect the principal and two contiguous lateral axes at their unit lengths. This necessitates each plane being parallel to the remaining lateral axis, as will be clear from the adjoining cross-section of the ground-form (Fig. 161). The parameter symbol of such a form must therefore be $a_1 : a_2 : \infty a_3 : c$; and its index symbol, $\{1\bar{1}01\}$. The indices of the twelve planes bounding the complete form, designated by their particular signs, are

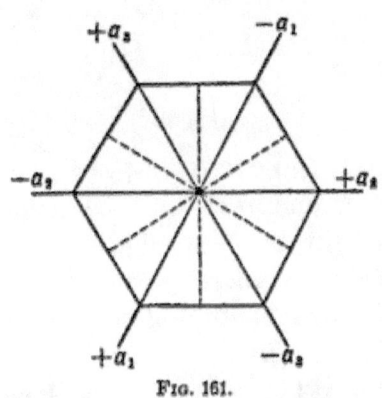
Fig. 161.

(above) $10\bar{1}1$ $01\bar{1}1$ $\bar{1}101$ $\bar{1}011$ $0\bar{1}11$ $1\bar{1}01$
(below) $10\bar{1}\bar{1}$ $01\bar{1}\bar{1}$ $\bar{1}10\bar{1}$ $\bar{1}01\bar{1}$ $0\bar{1}1\bar{1}$ $1\bar{1}0\bar{1}$

---

* Other authors have employed other sets of axes for the hexagonal system. Schrauf used only three, at right angles, the ratio between the two lateral axes being $1 : \sqrt{3}$. This he called the *orthohexagonal* system. Miller used three axes parallel to the edges of the fundamental rhombohedron.

THE HEXAGONAL SYSTEM.

The axial ratio for a given substance may be calculated from any hexagonal pyramid that is assumed as its ground-form, just as in the tetragonal system. The only difference between Fig. 162 and Fig. 125 (p. 84) is that the side a of the spherical triangle is 60° instead of 45°. The value of the axis c, in terms of the lateral axis a, is equal to tg b; while tg $b =$ tg $B \sqrt{\frac{3}{4}}$, or sin $b =$ cotg $A \sqrt{3}$, where B is one half the basal edge of the pyramid, and A one half its polar edge.

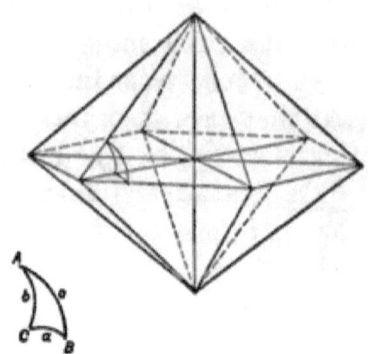

Fig. 162.

Development of the possible Holohedral Forms in the Hexagonal System. This is strictly analogous to the development of holohedral tetragonal forms (p. 85). The most general parameter formula is

$$na_1 : a_2 : pa_3 : mc;$$

but, inasmuch as the three lateral axes are fixed in their mutual inclinations at 60°, the intersection of two of them by any plane, at distances n and 1 from the centre, determines the point of intersection on the third axis by the same plane, as $\frac{n}{n-1}$ from the centre.* Hence the most general hexagonal formula becomes

$$na_1 : a_2 : \frac{n}{n-1} a_3 : mc.$$

---

* For proof of this see Klein's Einleitung in die Krystallberechnung, p. 319 (1876).

which, like the most general tetragonal formula, contains only the two variables m and n.

The limiting values for the vertical parameter m are zero and infinity, as in the tetragonal system (p. 86). If the shortest of the three lateral intercepts be assumed as unity and the intermediate one be designated by n, then the limiting values for the variable lateral parameters must be $n > 1$ and < 2, and $\frac{n}{n-1} > 2$ and $< \infty$. An inspection of Fig. 163 will make this clear. The shortest of any three finite intercepts for the same plane must be intermediate in its position between the other two. If we vary the position of any plane about its shortest intercept (here $- a_2$), so that its intercepts on the other axes receive different values; then, when $n = 2$, $\frac{n}{n-1} = 2$ also; if now n be diminished, $\frac{n}{n-1}$ will be proportionately increased, until when $n = 1$, $\frac{n}{n-1} = \infty$.

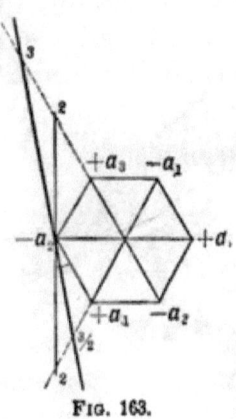

FIG. 163.

From the foregoing we see that, without the use of signs, only two permutations of the most general formula are possible:

$$na_1 : a_2 : \frac{n}{n-1} a_3 : mc \quad \text{and} \quad \frac{n}{n-1} a_1 : a_2 : na_3 : mc.$$

These correspond exactly to the two permutations of the most general formula possible in the tetragonal system (p. 85). They indicate that the most general hexagonal form contains but two planes in a dodecant, and that it is therefore bounded by twenty-four faces. This is also the number of secants into which all of the holohedral planes of symmetry divide space (p. 105).

By assigning limiting values to one or both of the parameters we find that there are seven types of hexagonal holohedrons which are entirely similar to those of the tetragonal system (p. 86).

Class I. Forms with *two* variable parameters.

1. $m \gtrless n$, general symbol becomes

$$na_1 : a_2 : \frac{n}{n-1} a_3 : mc.$$

Class II. Forms with *one* variable parameter.

2. $n = 1$, general symbol becomes $a_1 : a_2 : \infty a_3 : mc$.
3. $n = 2$, general symbol becomes $2a_1 : a_2 : 2a_3 : mc$.
4. $m = \infty$, general symbol becomes

$$na_1 : a_2 : \frac{n}{n-1} a_3 : \infty c.$$

Class III. Forms with *no* variable parameter.

5. $m = \infty$ and $n = 1$, general symbol becomes
$$a_1 : a_2 : \infty a_3 : \infty c.$$

6. $m = \infty$ and $n = 2$, general symbol becomes
$$2a_1 : a_2 : 2a_3 : \infty c.$$

7. $m = 0$ and $n = 1$, general symbol becomes
$$a_1 : a_2 : \infty a_3 : 0c.$$

The Dihexagonal Pyramid. The two planes of the most general hexagonal form belonging to each dodecant combine to form a double pyramid bounded by twenty-four similar scalene triangles. It is entirely analogous to the most general tetragonal form (p. 87), and is therefore called the *dihexagonal pyramid* (Fig. 164). This pyramid has three sorts of solid angles and three sorts of edges. Its polar edges must always be alternately dissimilar, because that particular case where they would be equal involves the irrational parameter sin 75°. $\sqrt{2} = 1.36666 +$, and is therefore crystallographically impossible.

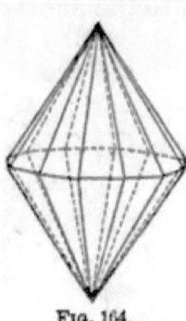

Fig. 164.

Inasmuch as one of the two variable lateral parameters, n and $\dfrac{n}{n-1}$, always determines the value of the other, it is only necessary to write one of them in the abbreviated symbol. The usage of Naumann is to write the *smaller* of the two, n, so that his symbol for the dihexagonal pyramid is mPn, like that for the ditetragonal pyramid, except that n can here never have a value greater than 2.

If we employ the designation of the hexagonal axes suggested by Bravais (p. 106), the most general expression for the indices of any plane becomes ($hikl$), in which the three first values refer to the lateral axes. At least one of these three lateral indices must in every case be negative, and their algebraic sum is always equal to zero, $h + i + k = 0$.*

There is a difference in the usage of different authors as to which particular index (the largest, medium, or smallest) a particular letter, h, i or k, represents. We

---

* For demonstration of this, see Groth's Physikalische Krystallographic, 2d Ed., p. 316.

shall follow the usage of Groth in designating the numerically largest index (corresponding to the shortest intercept) by the letter h, the medium index by k, and the smallest by i. Hence, without regard to signs, $2k > h > k > i$. (Fig. 165.)

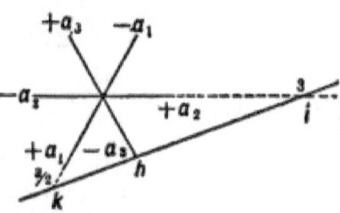

Fig. 165.

The *order* in which the indices are written in the symbol of a particular plane is always that given above for the axes, $a_1 : a_2 : a_3 : c$. Whichever of the lateral indices (h, i or k) refers to the axis a_1 must be written first, etc. Suppose, for example (Fig. 165), a plane cuts the three lateral axes with the parameters $\frac{3}{2}a_1 : 3a_2 : -a_3$; then its indices, or the reciprocals of the parameters, become $\frac{2}{3}, \frac{1}{3}, \bar{1}$ or $21\bar{3}$, which corresponds to the order $ki\bar{h}$.

The index l, referring to the vertical axis, is invariably written last; thus the general index symbol for the dihexagonal pyramid becomes $\{\bar{h}ikl\}$ or $\{h\bar{i}kl\}$, where $h = -(1+k)$; $n = \frac{h}{k}$ and $m = \frac{h}{l}$.

From what has just been stated it will be clear that the general index symbols for the twelve upper planes of a dihexagonal pyramid, commencing with the front dodecant and passing around to the left, must be

$ki\bar{h}l$ $h\bar{k}il$ $\bar{i}hkl$ $\bar{k}i\bar{h}l$ $\bar{h}kil$ $\bar{i}hkl$

$\bar{h}\bar{i}kl$ $\bar{k}hil$ $\bar{i}khl$ $hikl$ $kh\bar{i}l$ $ikhl$

The indices of the twelve lower planes will be the same with a negative sign over each l.

Hexagonal Pyramids of the First and Second Orders. These are obtained just as in the tetragonal system by giving the limiting values, 1 and 2, to the lateral parameter n of the general formula.

The first of these values yields the parameter symbol $a_1 : a_2 : \infty a_3 : mc$, which stands for a double hexagonal pyramid (Fig. 166). This form is bounded by twelve similar isosceles triangles, each of which occupies one dodecant and therefore corresponds to two contiguous faces of the most general form. There are two kinds of edges, basal and polar, and two kinds of solid angles. The lateral axes terminate in the solid angles, which determines this pyramid, like its corresponding tetragonal form (p. 88), to be of the *first order*. Its general abbreviated parameter and index symbols are mP and $\{h0\bar{h}l\}$. For the special case of a hexagonal pyramid of the first order whose vertical parameter is unity, we have the ground-form of the system (p. 106) whose symbols are P and $\{10\bar{1}1\}$.

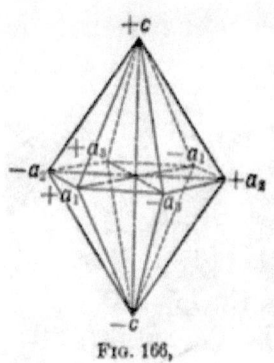

Fig. 166.

The assigning of the maximum limit to n produces the parameter symbol

$$2a_1 : a_2 : 2a_3 : mc,$$

which represents a hexagonal pyramid identical with that last described in all respects but position (Fig. 167). The lateral axes here terminate, not in the solid angles, but in the centers of the lateral or basal edges, which marks the form as one of the *second order*. Its symbols are $mP2$ and

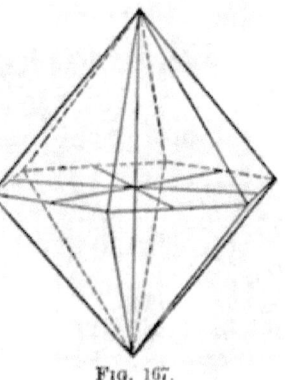

Fig. 167.

THE HEXAGONAL SYSTEM. 113

$\{kkhl\}$; or, for the special case where the vertical parameter is unity, $P2$ and $\{11\bar{2}2\}$.

The diagram (Fig. 168) is intended to exhibit the relative positions of the three hexagonal pyramids in cross-section. The inner hexagon corresponds to first-order forms, and the outer one to second-order forms, while the intermediate dodecagon represents the position of dihexagonal forms (cf. Fig. 129, p. 89). The values of the first- and second-order forms would, of course, be reversed if we chose to select the set of intermediate axes (dotted lines) as axes of reference.

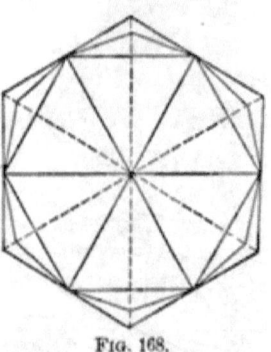

FIG. 168.

Hexagonal Prisms and Basal Pinacoid. These are so strictly analogous to the corresponding forms in the tetragonal system that they require but a word of explanation. The three prisms are derived from the three possible types of pyramids, by giving its maximum value, infinity, to the vertical parameter. (Figs. 169, 170, and 171.)

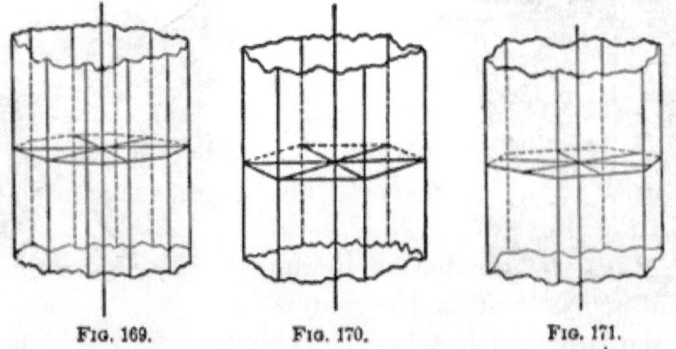

FIG. 169.   FIG. 170.   FIG. 171.

These forms are all open, i.e. do not of themselves enclose space, and are here represented as of indefinite extent, instead of in combination with the basal pinacoid, as are the corresponding figures (130, 131, and 132, p. 89) of the tetragonal system. The names and symbols of these prisms agree with those of the pyramids from which they are derived: *dihexagonal prism* (Fig. 169), ∞Pn, $\{h\bar{i}k0\}$; *hexagonal prism of the first order* (Fig. 170), ∞P, $\{10\bar{1}0\}$; *hexagonal prism of the second order* (Fig. 171), $\infty P2$, $\{2\bar{1}\bar{1}0\}$.

The *hexagonal basal pinacoid* (Fig. 172) is quite the same form as its tetragonal analogue (Fig. 133, p. 90). Its symbols are OP, $\{0001\}$.

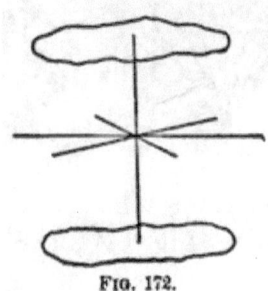

Fig. 172.

Limiting Forms in the Hexagonal System. The complete correspondence between the tetragonal and hexagonal systems is further illustrated by the following diagram of the limiting forms. It differs from that given on p. 91 only in the fact that the maximum limit for the lateral parameter is two instead of infinity.

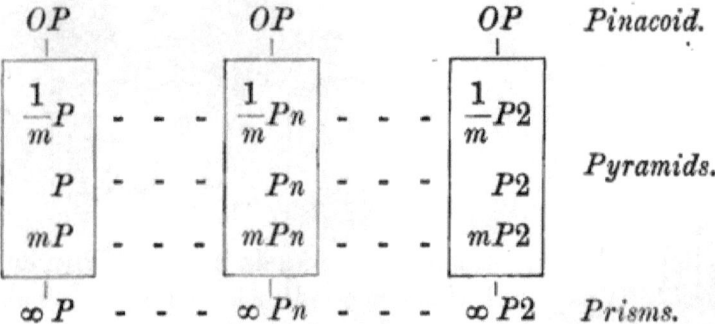

Holohedral Hexagonal Forms in Combination. The simpler hexagonal combinations are as readily intelli-

THE HEXAGONAL SYSTEM. 115

gible as those in the tetragonal system. Four of the seven possible holohedral types are open forms and therefore cannot occur uncombined.

The prisms may be closed by either the basal pinacoid (Fig. 173); by a pyramid (Fig. 174); or by both

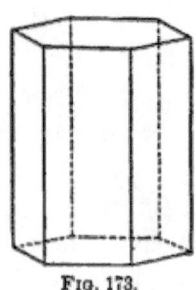
Fig. 173.

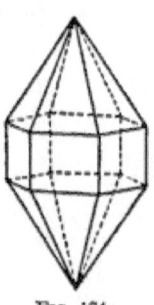
Fig. 174.

together. A hexagonal prism truncates the basal edges of a pyramid of the same order (Fig. 174), and the basal angles of a pyramid of the opposite order (Fig. 175). (Cf. Figs. 136 and 137, p. 93.) A pyramid

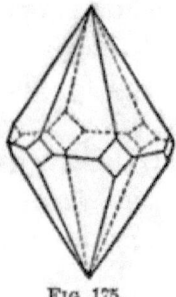
Fig. 175.

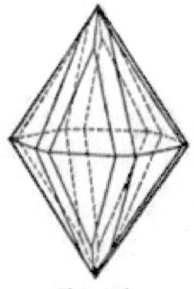
Fig. 176.

of the first order has its polar edges truncated by the faces of a pyramid of the second order whose vertical parameter is equal to its own (Fig. 176), just as in the tetragonal system (cf. Fig. 138, p. 94). The polar angles of any pyramid are bevelled by the faces of a more obtuse pyramid (Figs. 177 and 178); or its basal

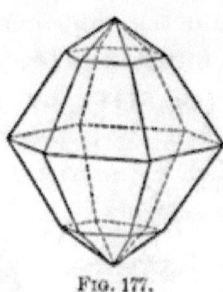
Fig. 177.

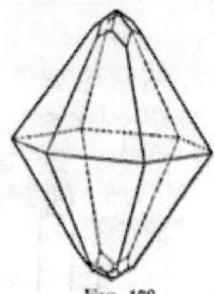
Fig. 178.

edges are bevelled by the planes of a more acute pyramid of the same order.

A more complex hexagonal combination is represented in Fig. 179, as it is sometimes observed on crystals of beryl. This shows the basal pinacoid, $0P$, $\{0001\}$ (c); the prism of the first order, ∞P, $\{10\bar{1}0\}$ (m); two pyramids of the first order, P, $\{10\bar{1}1\}$ (o) and $2P$, $\{20\bar{2}1\}$ (o^2); a pyramid of the second order, $2P2$, $\{11\bar{2}1\}$ (q); and a dihexagonal pyramid, $3P\frac{3}{2}$, $\{3\bar{2}\bar{1}1\}$ (s).

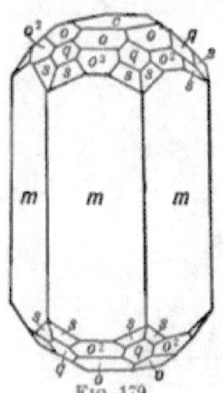
Fig. 179.

The number of substances known to exhibit holohedral hexagonal crystallization is very small. Among them may be mentioned the elements magnesium, beryllium, zinc, cadmium; the beryllium silicate, beryl (emerald); the antimono-silicate of manganese and iron, Langbanite; pyrrhotite (magnetic pyrites, Fe_7S_8); and, at high temperatures, tridymite, SiO_2.

Partial Hexagonal Forms. Partial crystal forms produced by hemihedrism, tetartohedrism, and hemimorphism attain their maximum importance in the hexagonal system. As has been just remarked, only a very small proportion of hexagonal substances possess the complete holohedral symmetry, and hence the various subdivisions of partial forms are here deserving of particular attention. One of these, indeed—the rhombohedral—has, on account of its extremely frequent occurrence, been regarded by many authors as a distinct system.

HEMIHEDRAL DIVISION OF THE HEXAGONAL SYSTEM.

Possible Kinds of Hexagonal Hemihedrism. There are three ways in which one half of the planes of the dihexagonal pyramid may be selected so as to satisfy the conditions of hemihedrism. These three ways correspond precisely with the three methods of hemihedral selection applied to the ditetragonal pyramid, as

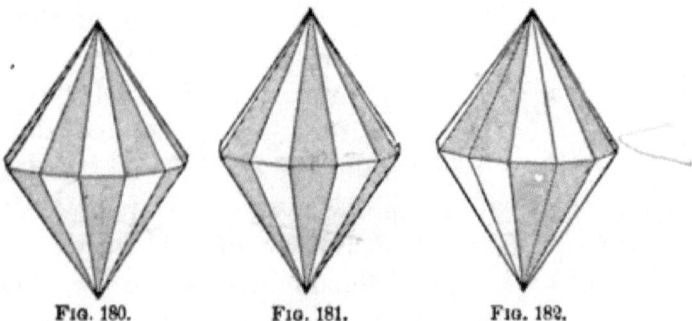

FIG. 180.   FIG. 181.   FIG. 182.

may be seen by comparing Figs. 180, 181, and 182 with Figs. 145, 146, and 147 on p. 96.

The first method of selection is by alternate planes of the most general form (Fig. 180). This produces new forms devoid of symmetry, and is called, as in the preceding system, *trapezohedral hemihedrism*. The second method of selection is by alternate pairs of planes intersecting in the basal edges (Fig. 181). It produces new forms with one principal, but no secondary planes of symmetry, and is called, as before, *pyramidal hemihedrism*. The third method of selection is by alternate dodecants (Fig. 182). This destroys all but the three intermediate secondary planes of symmetry, and is called *rhombohedral hemihedrism*.

Trapezohedral Hemihedrism. The extension of alternate planes of the dihexagonal pyramid until they intersect produces a new half-form bounded by twelve similar trapeziums. The two figures resulting from the two sets of alternating planes are devoid of sym-

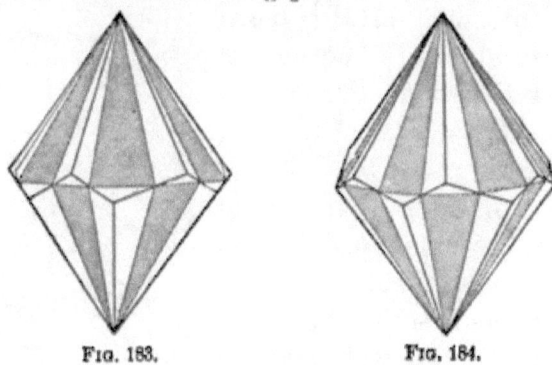

Fig. 183.   Fig. 184.

metry, and are therefore enantiomorphous (Figs. 183 and 184). They are called *hexagonal trapezohedrons* and are distinguished as right- and left-handed like their tetragonal analogues (p. 97). Their symbols are:

$$\frac{mPn}{2}r, \quad \tau\{ki\bar{h}l\}; \quad \text{and} \quad \frac{mPn}{2}l, \quad \tau\{\bar{h}ikl\}.$$

It is evident that none of the other hexagonal holohedrons can yield a geometrically new form by this method of selection, because each of their faces corresponds to at least two contiguous faces of the general form.

No example of trapezohedral hemihedrism has

been observed in the hexagonal system, with the possible exception of triethyl-trimesitate, $C_6H_3(CO_2.C_2H_5)_3$. Apparently holohedral crystals of this substance show slight variations in their alternating angles, which may be accounted for on the supposition that they are hexagonal trapezohedrons with very large indices. They may, however, also be regarded as pseudohexagonal crystals which very nearly approach hexagonal limiting forms.

Pyramidal Hemihedrism. The extension of the alternate pairs of planes on the dihexagonal pyramid which intersect in the basal edges, produces a hexagonal pyramid that differs from the holohedral pyramids of the first and second order only in its position. This is intermediate between the positions of the other two forms, since the axes terminate neither in the basal angles nor at the *centers* of the basal edges, but at some point in the latter, on one side of the center. This form is called the *hexagonal pyramid of the third order*. Its symbols are:

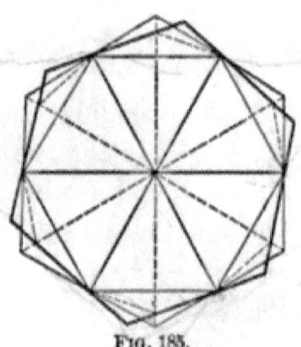

Fig. 185.

$$+\left[\frac{mPn}{2}\right], \quad \pi\{ki\bar{h}l\}; \quad \text{and} \quad -\left[\frac{mPn}{2}\right], \quad \eta\{h\overline{ikl}\}.$$

The position of this pyramid, relative to the two other hexagonal pyramids, is shown in the cross-section, Fig. 185, with which Fig. 168, p. 113, should be compared.

The planes of the dihexagonal prism correspond to the pairs of planes which alternately disappear from the dihexagonal pyramid in order to produce the pyramid

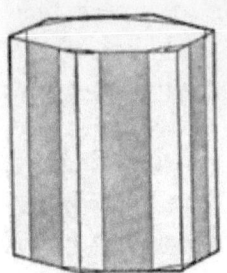

Fig. 186.

of the third order. This prism must therefore be capable of producing, by this method of selection, two corresponding *prisms of the third order* (Fig. 186) which are related to the prism of the first and second order in the same way that the pyramids of the third order are related to the other hexagonal pyramids (Fig. 185). The shortened parameter and index symbols of these prisms are:

$$+\left[\frac{\infty Pn}{2}\right], \quad \pi\{ki\bar{h}0\}; \quad \text{and} \quad -\left[\frac{\infty Pn}{2}\right], \quad \pi\{h\bar{i}k0\}.$$

The pyramidal hemihedrism is also parallel-face (p. 42), as in the tetragonal system. The forms of the third order have but one plane of symmetry, which is the horizontal or principal plane. The two half-forms derivable from the same holohedron are therefore congruent, and one may be brought into exactly the position of the other by a revolution about its vertical axis through an angle depending on the value of the lateral parameter, n.

None of the other hexagonal holohedrons are capable of producing geometrically new forms by the pyramidal hemihedrism, for the reasons already stated for the analogous tetragonal forms on p. 99. These will, however, be made clearer by an inspection of the four following figures (187, 188, 189, and 190), which represent the hexagonal forms of the first and second order with their faces

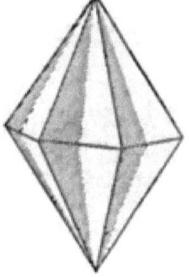

Fig. 187.

shaded to correspond to the pyramidal method of selection. On each figure a portion of every face

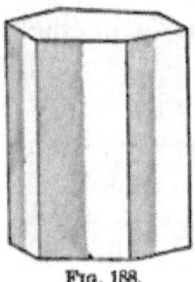
Fig. 188.

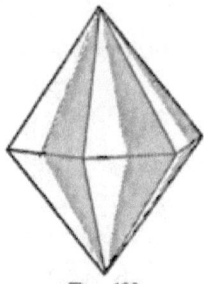
Fig. 189.

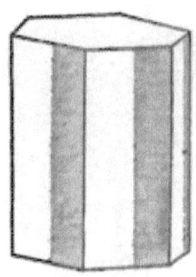
Fig. 190.

survives, which is enough by its extension to reproduce the form entire.

Fig. 191 represents a complex combination of hexagonal hemihedral forms produced by the pyramidal selection. They are observed on crystals of calcium phosphate, apatite ($Ca_5Cl(PO_4)_3$).

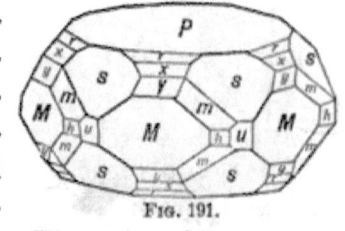
Fig. 191.

The nine forms comprise the basal pinacoid, $0P$, $\{0001\}$ (P); the prism of the first order, ∞P, $\{10\bar{1}0\}$ (M); prism of the second order, $\infty P2$, $\{2\bar{1}\bar{1}0\}$ (u), and prism of the third order, $-\left[\dfrac{\infty P\frac{3}{2}}{2}\right]$, $\pi\{3\bar{1}\bar{2}0\}$ (h); three pyramids of the first order, $\frac{1}{2}P$, $\{10\bar{1}2\}$ (r), P, $\{10\bar{1}1\}$ (x), and $2P$, $\{20\bar{2}1\}$ (y); one pyramid of the second order, $2P2$, $\{11\bar{2}1\}$ (s); and one of the third order, $-\left[\dfrac{3P\frac{3}{2}}{2}\right]$, $\pi\{3\bar{1}\bar{2}1\}$ (m).

As other examples of this same method of crystallization may be mentioned the three similarly constituted compounds, pyromorphite, $Pb_5Cl(PO_4)_3$; mimetesite, $Pb_5Cl(AsO_4)_3$; and vanadinite, $Pb_5Cl(VO_4)_3$.

Rhombohedral Hemihedrism. The selection of hexagonal planes by alternate dodecants can produce only two geometrically new forms, since on only two holohedrons do the faces belong exclusively to single dodecants.

The extension of the dihexagonal planes occupying alternate dodecants yields a new hemihedron, bounded by twelve similar scalene triangles, and called the *hexagonal scalenohedron*. The two corresponding half-forms possess the three intermediate planes of symmetry, and are consequently congruent. They have two sorts of polar edges, alternately more acute and obtuse; while the basal edges form a zigzag around each figure (Figs. 192 and 193). These two forms are distinguished as positive and negative, and their symbols are written:

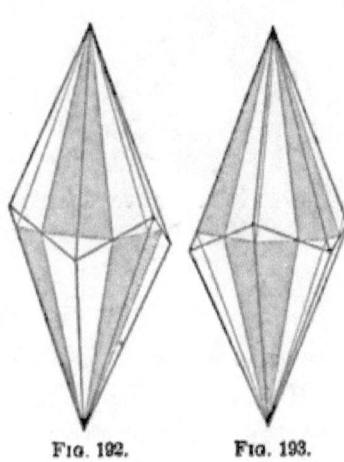

Fig. 192. Fig. 193.

$$+\frac{mPn}{2}, \quad \kappa\{h\bar{\imath}\bar{k}l\};* \quad \text{and} \quad -\frac{mPn}{2}, \quad \kappa\{ik\bar{h}l\}.$$

A scalenohedron whose polar edges are equal is crystallographically impossible, for the reason already given for the dihexagonal pyramid (p. 110).

---

* The Greek letter κ ($\kappa\lambda\iota\nu o\varsigma$, inclined) is retained to designate the indices of this hemihedrism in order to emphasize its analogy to the sphenoidal hemihedrism of the tetragonal system, although, as may be seen by an inspection of the figures, the new forms resulting in this case are not inclined-face, but parallel-face forms (cf. p. 42).

Each face of the hexagonal pyramid of the first order occupies one dodecant, and this form is therefore capable of producing rhombohedral hemihedrons. The extension of its two sets of alternating planes yields two new, congruent half-forms, which are each

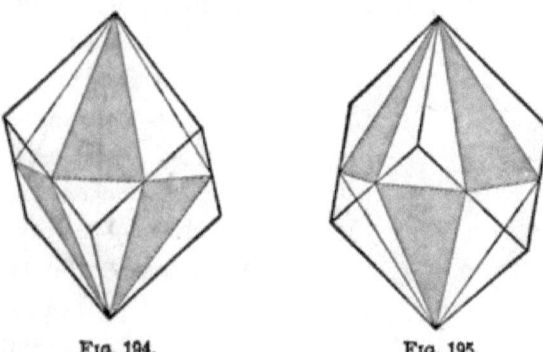

Fig. 194.   Fig. 195.

bounded by six similar rhombs and called *rhombohedrons* of the first order (Figs. 194 and 195). Their symmetry is the same as that of the scalenohedrons; and, like these, they become coincident by a revolution of either through 60° about its vertical axis. Their symbols are:

$$+\frac{mP}{2}, \quad \kappa\{h0\bar{h}l\}; \quad \text{and} \quad -\frac{mP}{2}, \quad \kappa\{0h\bar{h}l\}.$$

Rhombohedrons may have their polar edges more or less acute than their lateral edges. The first are known as *acute*, and the latter as *obtuse rhombohedrons*. The limiting form between these two kinds of rhombohedrons would have all of its edges similar, and would correspond to a hexagonal pyramid whose axial ratio is 1 : 1.2247. The interfacial angles on such a form would all be 90°, and it would therefore not differ geometrically from a cube.

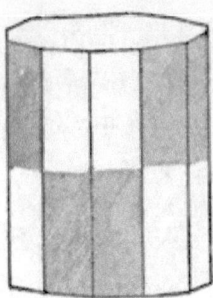

FIG. 196.

The four following figures (196–199) will make it evident that no other new types of forms can result by the rhombohedral selection of planes. Each of these holohedrons is shaded to correspond to a disappearance of the faces of alternate dodecants, and yet a portion of each holohedral plane will survive in every case, sufficient to reproduce the form as it was before.

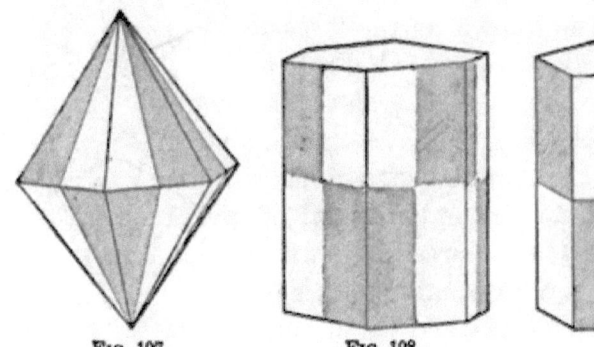

FIG. 197.   FIG. 198.   FIG. 199.

Abbreviated Symbols of Rhombohedrons and Scalenohedrons. Rhombohedrons and scalenohedrons are of such frequent occurrence on natural crystals that certain abbreviated symbols have been suggested for them by Naumann, which have come into general use.

Rhombohedrons are designated by the capital initial R, preceded by the vertical parameter of the hexagonal pyramid from which they are derived. Only negative rhombohedrons are distinguished by a sign; symbols without signs are to be considered as positive. For instance, R is the positive rhombohedron derived from the ground-form or fundamental pyramid, P. It is equivalent to the symbol $+\frac{P}{2}$. The rhombohedron $-8R$ is the negative form derived from the pyramid of the first order, $8P \cdot \frac{1}{4}R$ is the positive rhombohedron derived from $\frac{1}{4}P$, etc.

The shortened symbols of scalenohedrons are formed upon a different principle. In every scalenohedron it is possible to inscribe such a rhombohedron that the lateral edges of both forms shall exactly coincide (Fig. 200). This is known as the "*rhombohedron of the middle edges.*" A whole series of more or less acute scalenohedrons may have the same lateral edges, and hence may all be considered as derived from the same "rhombohedron of the middle edges" by increasing its vertical axis by different rational increments, and then joining the extremities of this axis to the lateral angles of the rhombohedron. Naumann's abbreviated symbol for a scalenohedron consists of the symbol of its "rhombohedron of the middle edges," $m'R$, followed by an index, n', to indicate by what quantity its semi-vertical axis is multiplied. The new symbol is therefore $m'R^{n'}$, in which m' and n' are quite different quantities from the two parameters, m and n, of the dihexagonal pyramid, of which the scalenohedron is a half-form $\left(\pm \frac{mPn}{2}\right)$.

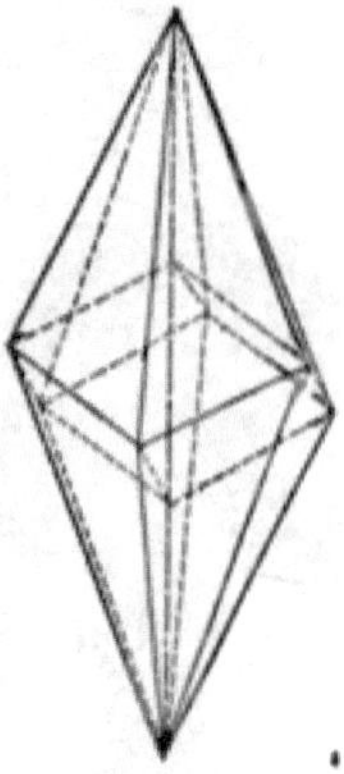

FIG. 200.

In order to be able to transform the abbreviated into the full symbols, and *vice versa*, we must know the relation existing between these four quantities: m', n' and m, n.

The parameter symbol of the "rhombohedron of middle edges" belonging to a scalenohedron derived from $na_1 : a_2 : \dfrac{n}{n-1}a_3 : mc$ is $a_1 : a_2 : \infty a_3 : \dfrac{m(2-n)}{n}c$; hence the symbol $m'R$, expressed in terms of the parameters of its corresponding scalenohedron, becomes $\dfrac{\dfrac{m(2-n)}{n}P}{2}$. Any particular one of the series of scalenohedrons having this same "rhombohedron of the middle edges" is designated by the index n', whose value depends upon m, the vertical parameter of its corresponding dihexagonal pyramid, $n' = \dfrac{m}{m'}$. To obtain the abbreviated symbol for a scalenohedron from its full symbol, $\pm \dfrac{mPn}{2}$, we have:

$$\frac{m(2-n)}{n} = m' \quad \text{and} \quad \frac{m}{m'} = n';$$

and hence, for the reverse transformation:

$$m'n' = m \quad \text{and} \quad \frac{2n'}{1+n'} = n.$$

The indices of the "rhombohedron of the middle edges" belonging to a scalenohedron, $\kappa\{h\bar{i}kl\}$, are $2k-h$, 0, $-2k+h$, l.* Hence, if the indices of the dihex-

---

* For proof of this see Groth's Physikalische Krystallographie, 2d ed., p. 342.

agonal pyramid are given, we have, for obtaining the abbreviated symbol of Naumann,

$$\frac{2k-h}{l}=m' \quad \text{and} \quad \frac{h}{2k-h}=n';$$

or, for the reverse transformation,

$$2n'=h; \quad n'+1=k; \quad n'-1=i; \quad \frac{2}{m'}=l.$$

Example. Given the dihexagonal pyramid, $4P\frac{4}{3}\{4\bar{1}\bar{3}1\}$, to find the shortened symbol for its scalenohedron.

$$\frac{4(2-\frac{1}{3})}{\frac{4}{3}}=2=m' \quad \text{and} \quad \frac{4}{2}=2=n'.$$

$2R^2$ is the symbol required. Using the indices, we obtain the same:

$$\frac{6-4}{1}=2=m' \quad \text{and} \quad \frac{4}{6-4}=2=n'. \quad 2R^2.^*$$

Rhombohedral Forms in Combination. Rhombohedral combinations differ from holohedral hexagonal combinations only when rhombohedrons or scalenohedrons are present. These forms are, however, so common that their more frequent combinations deserve mention.

A rhombohedron has its polar angles blunted by the faces of another rhombohedron of the same sign, the combination edges of the two being parallel (Fig. 201).

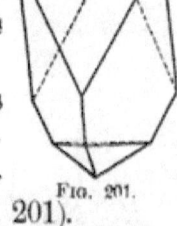

Fig. 201.

---

* Dana still further shortens Naumann's symbols by omitting the initial R except in the case of the fundamental rhombohedron. Thus $-\frac{1}{2}R$ becomes $-\frac{1}{2}$; $4R$, 4; R^3, 1^3; $-\frac{1}{2}R^5$, $-\frac{1}{2}^5$; etc. On Miller's method of designating rhombohedral forms, see Groth's Physikalische Krystallographie. 2d ed., p. 438.

Much more frequent are combinations of rhombohedrons of opposite signs. These generally show truncations of the polar edges of the more acute form, which can only be accomplished by a rhombohedron of the *opposite sign and half the vertical parameter*. For instance, R truncates the polar edges of $-2R$ (r), and is in turn truncated by the faces of $-\frac{1}{2}R(n)$, etc. (Fig. 202).

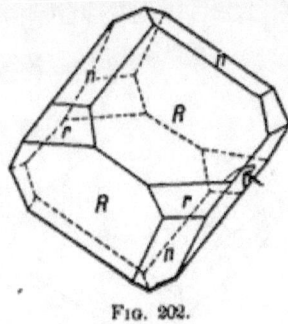

Fig. 202.

Figs. 203 and 204 show combinations of rhombohedrons of opposite signs where the above relation does not

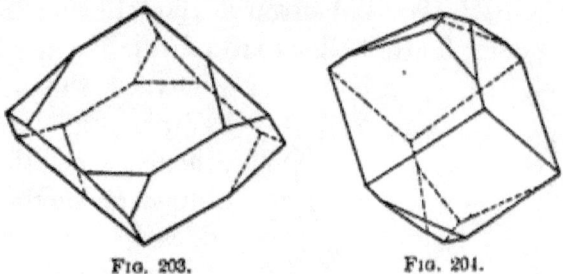

Fig. 203.   Fig. 204.

obtain. In the first case the forms have equal vertical parameters, R and $-R$; in the second, a rhombohedron

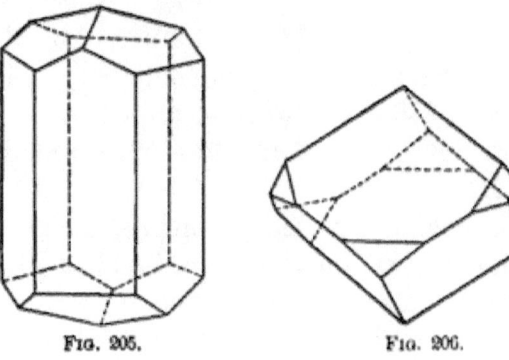

Fig. 205.   Fig. 206.

is modified by a form with *less* than half of its vertical parameter, R and $-\frac{1}{4}R$. Figs. 205 and 206 show a rhombohedron in combination with a prism of the first order; Figs. 207 and 208, the same form united to a prism of the second order.

Combinations of rhombohedrons and scalenohedrons are very manifold. A scalenohedron has its polar angles re-

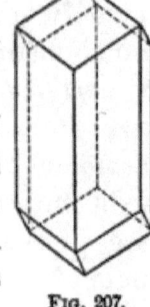

Fig. 207.

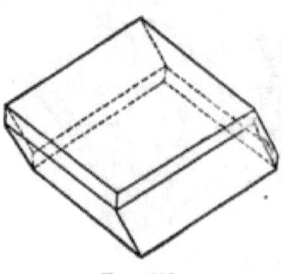

Fig. 208.

placed by its "rhombohedron of the middle edges" so that the combination edges are parallel to the lateral edges of the scaleno-hedron (Fig. 209). In other cases either the obtuse or acute polar edges of the scaleno-hedron are truncated by the planes of a rhombohedron (Fig. 210); or the polar edges of a rhombohedron are

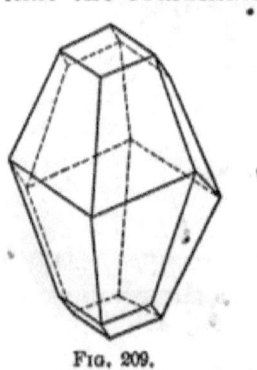

Fig. 209.

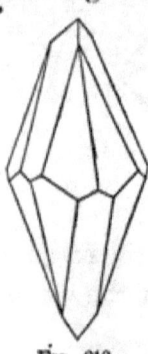

Fig. 210.

bevelled by the faces of a scalenohedron (Fig. 211). Figs. 212 and 213 show the scalenohedron in combination with the prisms of the first and second orders respectively.

The rhombohedral hemihedrism is of such very common occurrence that only a few of its most prominent representatives need be mentioned here. Examples of somewhat complex combinations of rhombohedral

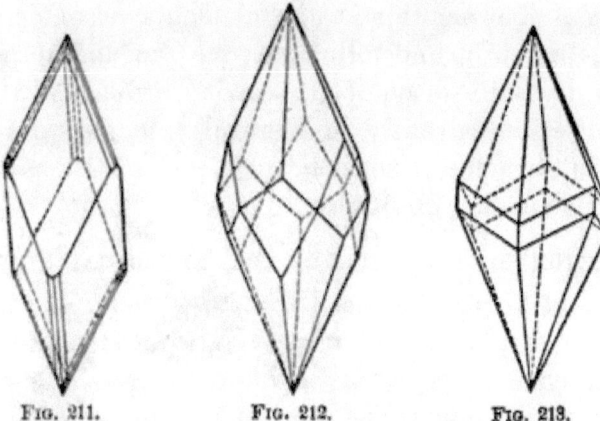

Fig. 211. Fig. 212. Fig. 213.

forms are given in the two following figures. Fig. 214 represents a crystal of hematite (Fe_2O_3) bounded

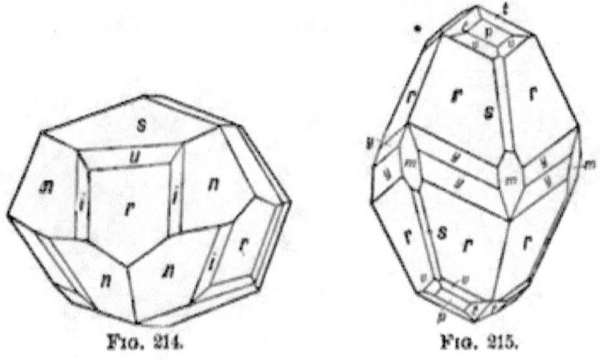

Fig. 214. Fig. 215.

by the forms R, $\kappa\{10\bar{1}1\}(r)$; $\tfrac{2}{3}R$, $\kappa\{30\bar{3}5\}(u)$; $\tfrac{1}{4}R$, $\kappa\{10\bar{1}4\}(s)$; $\tfrac{4}{3}P2$, $\kappa\{22\bar{4}3\}(n)$; and $\tfrac{2}{5}R^2$, $\kappa\{42\bar{6}5\}(i)$. Fig. 215 shows a crystal of calcium carbonate (calcite) with the forms R, $\kappa\{10\bar{1}1\}(p)$; $\tfrac{5}{2}R$, $\kappa\{50\bar{5}2\}(s)$; $4R$, $\kappa\{40\bar{4}1\}(m)$; R^2, $\kappa\{3\bar{1}\bar{2}1\}(r)$; R^3, $\kappa\{5\bar{2}\bar{3}1\}(y)$; $R^{\frac{5}{3}}$, $\kappa\{7\bar{1}\bar{6}5\}(u)$; $\tfrac{1}{4}R^2$, $\kappa\{3\bar{1}\bar{2}4\}(t)$. This mineral exhibits a greater variety of forms than any other rhombohedral substance.

Among other examples of this mode of crystalliza-

THE HEXAGONAL SYSTEM.

tion may be mentioned the elements arsenic, antimony, bismuth, and tellurium; ice; corundum, Al$_2$O$_3$ (generally with forms of the second order, and therefore often apparently holohedral); magnesium hydroxide (brucite); sodium nitrate (NaNO$_3$) and the hydrous silicate, chabazite.

TETARTOHEDRAL DIVISION OF THE HEXAGONAL SYSTEM.

Kinds of Tetartohedrism. Tetartohedrism, like hemihedrism, attains its maximum importance in the hexagonal system. We may develop the possible kinds of hexagonal tetartohedrism by examining the effect of a simultaneous application of two different kinds of hemihedrism to the planes of the most general holo-

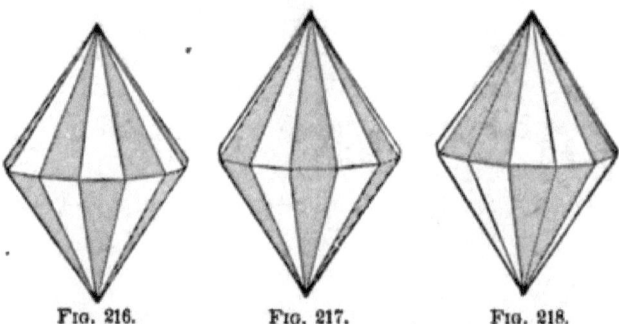

FIG. 216.   FIG. 217.   FIG. 218.

hedral form, as was done in the preceding systems. To accomplish this we may imagine each of the three adjoining figures (216, 217 and 218) laid upon one of the others and note the result. A union of the third with either the first or second produces a survival of three planes above and three others below which satisfy the conditions of tetartohedrism; while the second superposed upon the first leaves six planes on the lower half of the crystal and none on the upper half.

In order to bring out this result still more clearly, we may number the planes of the dihexagonal pyramid as is indicated in Fig. 219, and then write out the numbers representing the twenty-four planes in two rows, as follows:

(above) 1 2 3 4 5 6 7 8 9 10 11 12
(below) 1 2 3 4 5 6 7 8 9 10 11 12

If now we erase by a mark to the right those planes which would disappear by one kind of hexagonal hemihedrism, and by a mark to the left those planes which would disappear by another kind of hemihedrism, we can obtain the three following results:

1. We may combine the trapezohedral and rhombohedral hemihedrisms as follows:

(above) 1̸ 2 3̸ 4̸ 5̸ 6 7̸ 8 9̸ 10 11̸ 12̸
(below) 1̸ 2̸ 3 4̸ 5̸ 6̸ 7 8 9̸ 10̸ 11 12̸

There remain (above) .. 2 ... 6 ... 10 ...
(below) .... 3 ... 7 ... 11 .

This selection yields a possible tetartohedrism which is called the *trapezohedral*.

2. We may combine in the same manner the pyramidal and rhombohedral hemidrisms as follows:

(above) 1̸ 2 3̸ 4̸ 5 6 7̸ 8 9̸ 10 11̸ 12
(below) 1̸ 2 3̸ 4 5̸ 6̸ 7̸ 8 9̸ 10 11̸ 12

There remain (above) .. 2 ... 6 ... 10 ...
(below) .... 4 ... 8 12

giving another possible tetartohedrism which is called the *rhombohedral*.

3. Finally, we may combine the trapezohedral and pyramidal hemihedrisms:

(above) 1̸ 2 3̸ 4 5̸ 6 7̸ 8 9̸ 10 11̸ 12
(below) 1̸ 2̸ 3̸ 4̸ 5̸ 6̸ 7̸ 8̸ 9̸ 10̸ 11 12̸

There remain (above) .. 2 ... 4 ... 6 ... 8 ... 10 ... 12
(below) ....

This selection does not satisfy the conditions of tetartohedrism, but produces a *hemimorphism* in the direction of a vertical axis.

Trapezohedral Tetartohedrism. The simultaneous application of the trapezohedral and rhombohedral hemihedral selections to the faces of the dihexagonal pyramid produces the effect shown in Fig. 219. The

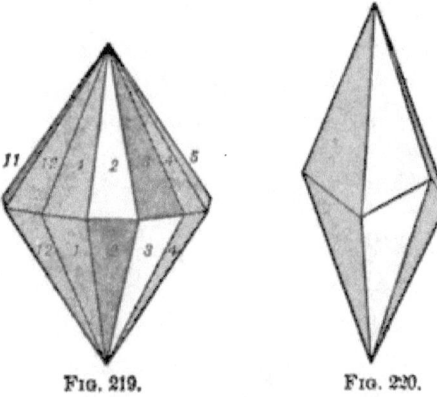

FIG. 219.   FIG. 220.

extension of the six surviving (white) planes gives an asymmetric solid, bounded by six similar trapeziums. The same result is secured by selecting one-half of the faces of the hexagonal trapezohedron by the rhombohedral; or one half of the faces of the scalenohedron by the trapezohedral method (Fig. 220). The four trapezohedral quarter-forms of the dihexagonal pyramid form two enantiomorphous pairs (Figs. 221 and 222), the members of each pair being themselves congruent. They are called *trigonal trapezoedrons*, and their symbols are written:

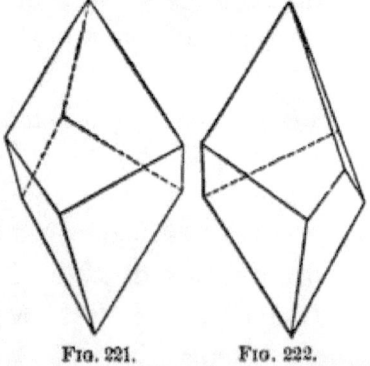

FIG. 221.   FIG. 222.

Positive right-handed,

$$+\frac{mPn}{4}r, \text{ or } \frac{mR^n}{2}r, \quad \kappa\tau\{ki\bar{h}l\}$$

Negative right-handed,

$$-\frac{mPn}{4}r, \text{ or } -\frac{mR^n}{2}r, \quad \kappa\tau\{h\bar{k}il\}$$

} congruent pair.

Positive left-handed,

$$+\frac{mPn}{4}l, \text{ or } \frac{mR^n}{2}l, \quad \kappa\tau\{\bar{h}i\bar{k}l\}$$

Negative left-handed,

$$-\frac{mPn}{4}l, \text{ or } -\frac{mR^n}{2}l, \quad \kappa\tau\{k\bar{h}il\}$$

} congruent pair.

The union of a right- and left-handed form would produce a scalenohedron; the union of two right-handed or two left-handed forms, a hexagonal trapezohedron.

The survival and extension of the corresponding portions of the planes of the dihexagonal prism produces a new prismatic form, bounded by six planes, intersecting in edges which are alternately more obtuse and more acute. (Fig. 223). This is called the *ditrigonal prism*. Two corresponding forms of this kind are derivable from each dihexagonal prism, which differ only in position, and whose symbols are

$$\frac{\infty Pn}{4}r, \quad \kappa\tau\{ki\bar{h}0\}; \quad \text{and} \quad \frac{\infty Pn}{4}l, \quad \kappa\tau\{\bar{h}i\bar{k}0\}.$$

Fig. 223.

The hexagonal pyramid of the second order, when subjected to analogous selection, retains a portion of its three alternating upper planes, as well as a portion of those planes directly below them (Fig. 224). The extension of these planes produces a solid bounded by six similar isosceles triangles, intersecting in *horizontal* basal edges. This form is called

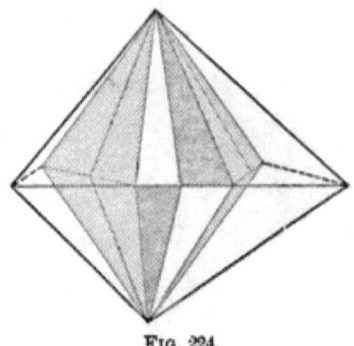

Fig. 224.

the *trigonal pyramid*. There is a congruent pair of them corresponding to the alternating sets of pyramidal planes, their symbols being:

$$\frac{mP2}{4}r, \quad \kappa\tau\{kk\bar{h}l\}; \quad \text{and} \quad \frac{mP2}{4}l, \quad \kappa\tau\{h\bar{k}\bar{k}l\}.$$

The hexagonal prism of the second order is only a special case of the pyramid of the second order, where the basal edges have become 180°. It must therefore yield two new prismatic forms, analogous to those last described, by the survival of its alternate planes (Fig. 225). These are called *trigonal prisms*, and are designated by the symbols

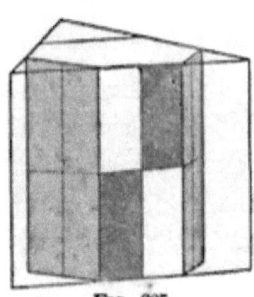

Fig. 225.

$$\frac{\infty P2}{4}r, \quad \kappa\tau\{11\bar{2}0\}; \quad \text{and} \quad \frac{\infty P2}{4}l, \quad \kappa\tau\{2\bar{1}\bar{1}0\}.$$

The same method of selection, applied to the hexagonal pyramid of the first order, yields a positive and a negative rhombohedron, which are not further modified in form by becoming tetartohedral (Figs. 226 and 227).

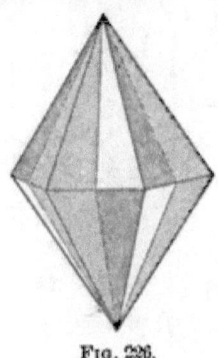

Fig. 226.

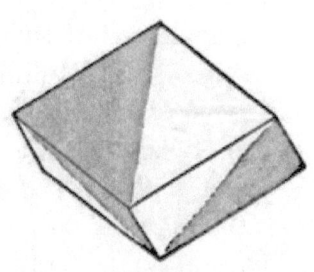

Fig. 227.

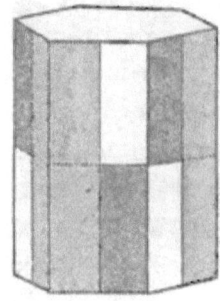

Fig. 228.

The hexagonal prism of the first order suffers no geometrical change whatever on becoming tetartohedral, as may be seen from Fig. 228; and the same is true of the only remaining holohedral form, the basal pinacoid.

Trapezohedral tetartohedral forms may also be regarded as the product of hemimorphism in the direction of the three lateral axes, and, like other hemimorphic crystals, they are pyroelectric. Their asymmetric molecular structure, and consequent enantiomorphism, requires that they should exhibit circular polarization, and this is, in fact, the case. The most prominent example of this mode of crystallization is offered by silica, SiO_2 (quartz). A not unusual combination of forms on crystals of this substance is given in Fig. 229. It shows the prism of the first order (m) unchanged; positive and negative rhombohedrons (r and r') apparently only hemihedral; while the trigonal pyramid (s) and the trigonal trapezohedron (x) occur as true quarter-forms.

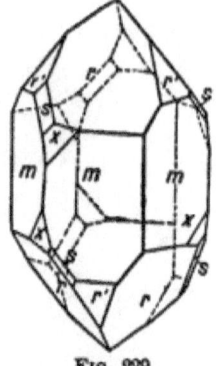

Fig. 229.

Other examples of this crystallization are: mercuric sulphide, HgS (cinnabar); the dithionates of potassium ($K_2S_2O_6$), of calcium ($CaS_2O_6 + 4$ aq), of strontium ($SrS_2O_6 + 4$ aq), of barium ($BaS_2O_6 + 4$ aq), and of lead ($PbS_2O_6 + 4$ aq); sodium-periodate ($NaIO_4 + 3$ aq); benzil ($C_{14}H_{10}O_2$); and matico-stearopten ($C_{10}H_{16}O$).

Rhombohedral Tetartohedrism. A combination of the pyramidal and rhombohedral methods of hemihedral selection (Figs. 217 and 218, p. 131) results in the survival of three planes in the upper half of the dihexagonal pyramid and of three other planes in the lower half.

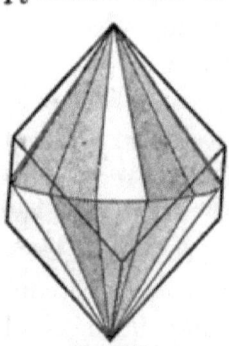

Fig. 230.

These are so distributed as to produce, by their intersections, a figure which does not differ from a hemihedral rhombohedron except in its position with reference to the crystallographic axes (Fig. 230). Each lateral axis terminates on one side of each of the faces, and the form is called the *rhombohedron of the third order*. The four quarter-forms derivable from each dihexagonal pyramid are designated by the following parameter and index symbols:

$$+ \frac{mPn}{4} \cdot \frac{r}{l}, \qquad \pi\kappa\{ki\bar{h}l\};$$

$$-\frac{mPn}{4}\cdot\frac{r}{\bar{l}}, \qquad \pi\kappa\{\bar{i}hk\bar{l}\};$$

$$+\frac{mPn}{4}\cdot\frac{l}{r}, \qquad \pi\kappa\{h\bar{i}\bar{k}l\};$$

$$-\frac{mPn}{4}\cdot\frac{l}{r}, \qquad \pi\kappa\{ik\bar{h}l\}.$$

These four quarter-forms possess three planes of symmetry, like other rhombohedrons, and are therefore congruent. The position of these planes of symmetry is, however, different from that of any of the holohedral hexagonal planes of symmetry (see Fig. 159).

This same method of selection, applied to the hexagonal pyramid of the second order, is capable of producing two congruent rhombohedrons which again differ from hemihedral rhombohedrons only in their positions (Fig. 231). These are called *rhombohedrons of the second order*, and they stand with reference to the intermediate axes (p. 105) just as the hemihedral rhombohedrons do with reference to the lateral axes of reference. They may therefore be made to coincide with the latter by a revolution about their vertical axes of 30°. Their symbols are:

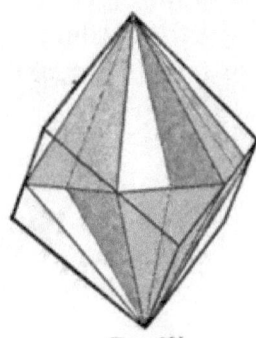

Fig. 231.

$$\pm\frac{mP2}{4}\cdot\frac{r}{l}, \qquad \pi\kappa\{kk\bar{h}l\};$$

$$\pm\frac{mP2}{4}\cdot\frac{l}{r}, \qquad \pi\kappa\{hkkl\}.$$

$$\pm\frac{mP2}{4}\cdot\frac{l}{r}$$

The same method of selection, if applied to the faces of the hexagonal pyramid of the first order, produces the same effect as the rhombohedral hemihedrism, i.e. two *rhombohedrons of the first order* (Fig. 232). Their symbols may be written:

$$\pm \frac{mP}{4} \cdot \frac{r}{l}, \qquad \pi\kappa\{0h\bar{h}l\};$$

$$\pm \frac{mP}{4} \cdot \frac{l}{r}, \qquad \pi\kappa\{h0\bar{h}l\}.$$

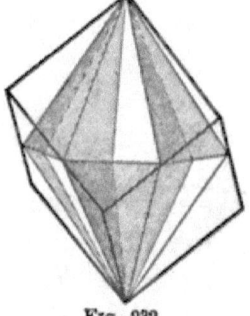

Fig. 232.

The relative positions of the three orders of rhombohedrons is well illustrated in the adjoining linear projection (Fig. 233), where those of the first order are drawn in heavy, those of the second order in faint, and those of the third order in dotted lines.

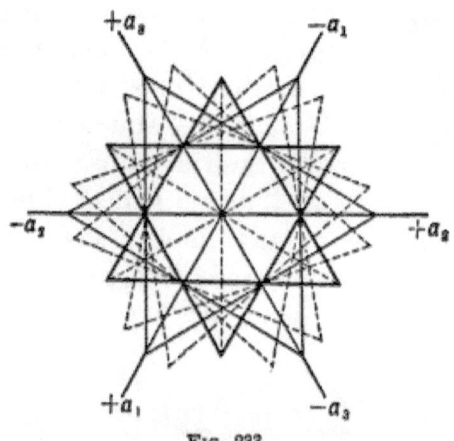

Fig. 233.

The other hexagonal holohedrons yield in this tetartohedrism no new forms. The dihexagonal prism produces two hexagonal prisms of the third order, just as it does in the pyramidal hemihedrism. The other two hexagonal prisms retain all their six faces.

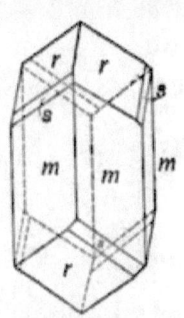

Fig. 234.

Examples of rhombohedral tetartohedrism in the hexagonal system are not rare. Fig. 234 shows a crystal of copper silicate (dioptase) with a rhombohedron of the third order, $-2R^{\frac{7}{3}}$, $\pi\kappa\{1\cdot 13\cdot\overline{14}\cdot 6\}$ (s), in combination with the prism of the second order, $\infty P2$, $\{2\overline{1}\overline{1}0\}$ (m), and the negative rhombohedron $-2R$, $\{02\overline{2}1\}$ (r).

Fig. 235 represents a more complicated crystal of the beryllium silicate, phenacite, which is mainly terminated by a rhombohedron of the third order, $-\dfrac{\frac{2}{3}P\frac{3}{2}}{4}r$, $\pi\kappa\{\overline{1}3\overline{2}2\}$ (x). With this form are associated the prisms of the first and the second order, ∞P, $\{10\overline{1}0\}$ (m), and $\infty P2$ $\{11\overline{2}0\}$ (a); another rhombohedron of the third order, $\dfrac{3P\frac{3}{2}}{4}r$, $\pi\kappa\{21\overline{3}1\}$ (s), and the two rhombohedrons of the first order, R, $\kappa\{10\overline{1}1\}$ (r), and $-\frac{1}{2}R$, $\kappa\{01\overline{1}2\}$. With this substance is isomorphous the corresponding zinc salt (willemite). Still other examples are the carbonates, magnesite, $MgCO_3$, and dolomite, $(Ca,Mg)CO_3$; possibly ilmenite, $FeTiO_3$, and some other substances.

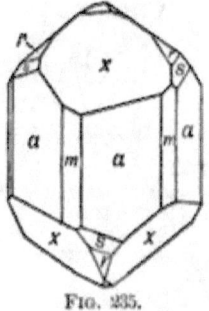

Fig. 235.

Hemimorphism. Hemimorphism in the direction of the vertical axis (p. 42) is particularly common in the hexagonal system. It may be considered as produced by a combination of the trapezohedral and pyramidal hemihedrisms (p. 133); and, as it generally occurs on crystals which are also rhombohedral, these may be viewed as exhibiting simultaneously all three hemihedrisms of the hexagonal system. Examples of such crystallization are offered by the sulphides of zinc, ZnS (wurtzite), and cadmium, CdS (greenockite); the sulpharsenide and sulphantimonide of silver, Ag_3AsS_3 (proustite, ruby silver) and Ag_3SbS_3 (pyrargyrite); the silicate tourmaline; and certain artificial salts. A combination observed on tourmaline is represented in Fig. 236. Its forms are the prisms ∞P, $\{10\bar{1}0\}$ (g), which, as a result of hemimorphism,

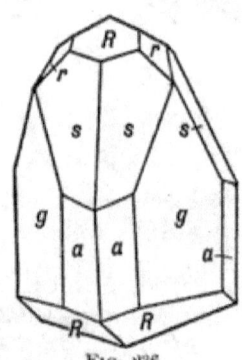

Fig. 236.

appears as a trigonal prism, and $\infty P2$, $\{11\bar{2}0\}$ (a); the scalenohedron R^3, $\kappa\{21\bar{3}1\}$ (s), and the two rhombohedrons R, $\kappa\{10\bar{1}1\}$ (R) and $-2R$, $\kappa\{02\bar{2}1\}$ (r), at the antilogue pole, while only one of these forms, R, occurs at the analogue pole.

Hemimorphism in the direction of the vertical axis produces the same result on holohedral, trapezohedral and pyramidal hemihedral crystals. Fig. 237 shows the hemimorphic development of the holohedral zinc oxide, ZnO (zincite), from Stirling Hill, N. J. The etched figures observed on the apparently holohedral silicate, nepheline, indicate a similar mode of crystallization, disguised by complicated twinning.

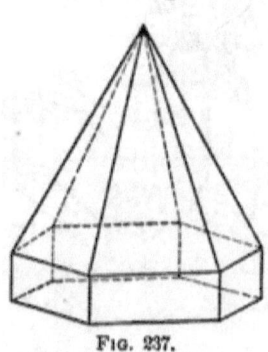

Fig. 237.

CHAPTER VI.

THE ORTHORHOMBIC SYSTEM.*

HOLOHEDRAL DIVISION.

Third Class of Crystal Systems. According to the classification of the crystal systems given on pages 44 and 45, all those whose complete forms possess no principal axis or plane of symmetry form the third or anisometric class. These systems are three in number, and they are united by a variety of common features, among which is their optically biaxial character and the fact that their planes are all referred to three unequal and therefore not interchangeable axes. The names of these three systems are the Orthorhombic, the Monoclinic, and the Triclinic, each of which forms the subject of one of the three following chapters.

Symmetry of the Orthorhombic System. The highest grade of symmetry possessed by any system of the third class is that of the orthorhombic system, whose holohedral forms have three secondary planes of symmetry at right angles to one another (Fig. 238). These planes therefore divide space into eight simi-

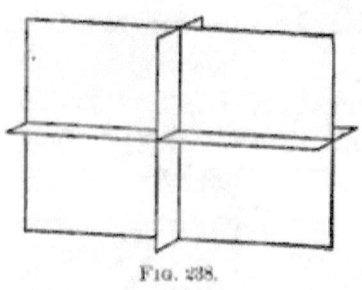
FIG. 238.

---

* Also called the *rhombic, prismatic,* and *trimetric* system.

THE ORTHORHOMBIC SYSTEM.

lar octants, like the axial planes of the isometric or tetragonal systems.

Axes. The orthorhombic axes of reference have their directions fully determined by the symmetry of the system. They are all axes of symmetry and must therefore be perpendicular to one another; but, because they lie in secondary planes of symmetry, no two of them are interchangeable, and hence they must all three be of unequal length. Since there is no principal axis of symmetry, any one of the three directions may be made the vertical axis, and there is in reality a great difference in the usage of different authors in this respect, even in regard to crystals of the same substance. It is customary to place the orthorhombic crystal in such a position that the longer of its two lateral axes will run from right to left. This is therefore called the *macrodiagonal*, while the shorter lateral axis, which runs from back to front, is known as the *brachydiagonal*. The letters representing the different axes are further distinguished by signs written over them; thus, a short sign over the brachydiagonal ($\breve{a}$), a long sign over the macrodiagonal ($\bar{b}$), and a perpendicular mark over the vertical axis ($\dot{c}$) (Fig. 239).

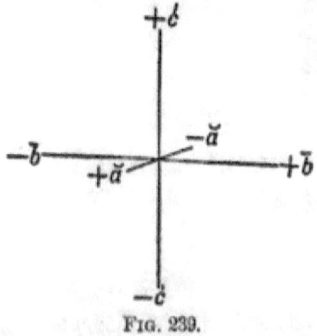

FIG. 239.

The distribution of the positive and negative extremities of the axes is the same as in the isometric and tetragonal systems.

Fundamental Form and Axial Ratio. No single orthorhombic crystal form (p. 35) can be bounded by more than eight planes; because, if none of the three axes are interchangeable, no permutations of the general parameter symbol are possible, and hence only one plane belongs to an octant. All orthorhombic pyramids are therefore alike in the number and distribution of their planes, which are always similar scalene triangles. It is possible to choose *any* pyramid which occurs on an orthorhombic crystal as the ground-form, but it is customary to select for this purpose the most frequent or prominent pyramid, or the one which will yield the simplest indices for the other planes.

When the choice of some particular pyramid as a ground-form has been made, the axial ratio of the substance to which it belongs may be calculated from it. The inequality in the lengths of all three axes produces a double ratio, in which the length of the macrodiagonal ($\bar{b}$) is taken as the unit. The nature of this axial ratio is the same as in the tetragonal system, except that in place of a single irrational quotient, $\frac{c}{a}$ (p. 83), we now have *two* such quotients to determine for each orthorhombic substance, $\frac{a}{b}$ and $\frac{c}{b}$. These quotients are called the *crystallographic constants* of the orthorhombic system. They fix the axial ratio, $\breve{a} : \bar{b} : \acute{c}$, and may be determined from the angles of the pyramid selected as the fundamental form as follows (Fig. 240): In the spherical triangle abc,

$$\cos a = \frac{\cos \tfrac{1}{2} Y}{\sin \tfrac{1}{2} Z}; \quad tg . a = \breve{a}.$$

$$\cos b = \frac{\cos \tfrac{1}{2} Z}{\sin \tfrac{1}{2} Y}; \quad tg . b = \acute{c}.$$

One axial ratio forms the basis of each orthorhombic crystal series (p. 92); and hence one such series belongs to each orthorhombic substance. No matter

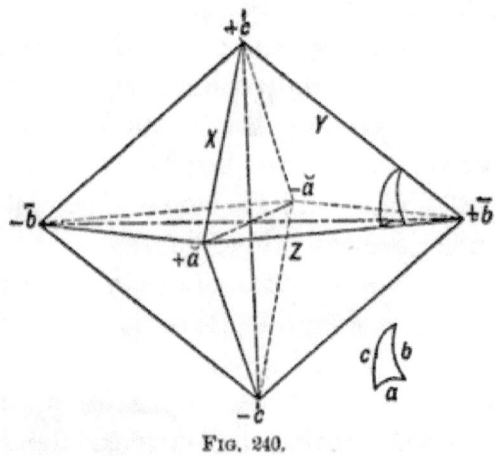

Fig. 240.

what pyramid is selected as the fundamental form, all the intercepts on the brachydiagonal axis of planes belonging to the same series must be rational multiples of its quotient, $\dfrac{\breve{a}}{\bar{b}}$; and all intercepts of planes on the vertical axis must be rational multiples of its quotient, $\dfrac{\acute{c}}{\bar{b}}$.

Derivation of the Holohedral Orthorhombic Forms. Since the most general orthorhombic symbol, $n\breve{a} : \bar{b} : m\acute{c}$, is capable of no permutations, it can only represent an eight-sided pyramid, essentially like the fundamental form (Fig. 240). This formula must stand for a different form from the formula $\breve{a} : n\bar{b} : m\acute{c}$, since $\breve{a}$ and $\bar{b}$ are not interchangeable. It is customary always to make the lesser of the two lateral parameters unity, so that the limiting values for n are one and infinity, and for *m* zero and infinity, as in the tetragonal system. Since, however, the parameter *n* may refer to either of the two dissimilar lateral axes, a considerable variety of form-types may result, which can best be considered under the three groups of pyramids, prisms, and pinacoids, defined on p. 36.

Pyramids. These are forms whose planes intersect all three axes. All orthorhombic pyramids are bounded by eight similar scalene triangles, which meet in three kinds of edges and in three kinds of solid angles (Fig. 240). It is, however, usual to distinguish three sorts of such pyramids, according to the lateral axis to which the parameter *n* refers.

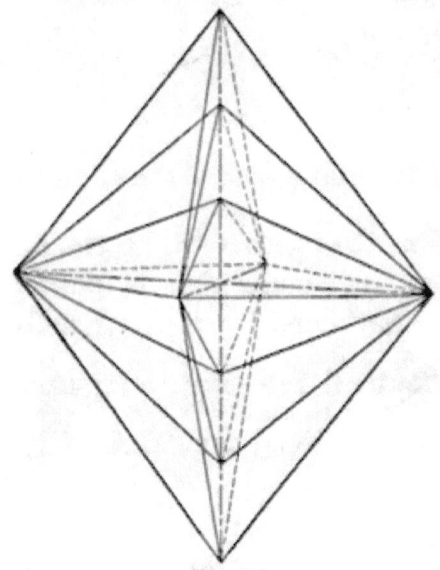

Fig. 241.

There is one series or zone* of pyramids both of whose lateral parameters are equal to unity, $\breve{a} : \bar{b} : m\breve{c}$. To this series the fundamental form belongs, and it is therefore called the zone of *unit pyramids*. Its limiting forms are the basal pinacoid, where $m = 0$, and the unit prism where $m = \infty$ (Fig. 241). The general parameter and index symbols of these pyramids are mP, $\{hhl\}$.

On one side of the zone of unit pyramids lie those whose lateral parameter, $n \, (> 1)$, refers to the macro-

---

* For the full explanation of this term, see Appendix, p. 217.

diagonal axis, $\bar{b}$. These are called *macropyramids*, and their general symbols are written $\breve{a} : n\bar{b} : m\dot{c}$, $mP\bar{n}$, $\{hkl\}$ $(h>k)$ (Fig. 242).

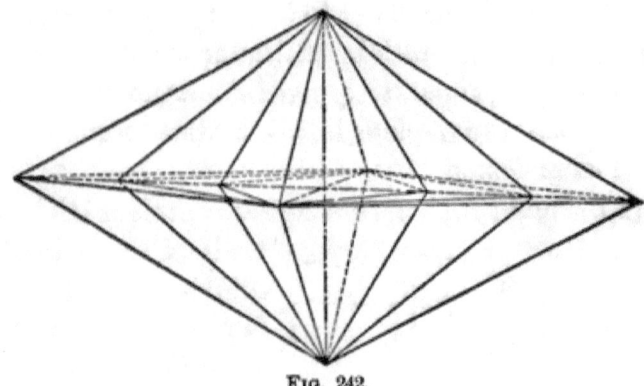

Fig. 242.

On the opposite side of the unit pyramids lies a third class of these forms whose lateral parameter, n, refers to the brachydiagonal axis, $\breve{a}$. These are called *brachypyramids*, and their general symbols are written: $n\breve{a} : \bar{b} : m\dot{c}$, $mP\breve{n}$, $\{hkl\}$ $(h<k)$ (Fig. 243).

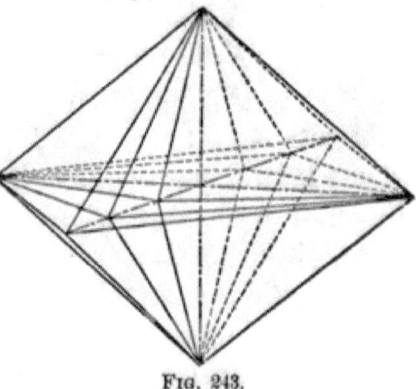

Fig. 243.

For every possible value of the lateral parameter, n, there is a vertical zone of macro- or brachypyramids which is limited by the basal pinacoid and a prism, as is the zone of unit pyramids.

Prisms. These embrace all forms whose planes are parallel to one axis, and which must therefore always have one sign of infinity in their parameter symbol, or one zero in their index symbol. All orthorhombic prisms are open forms (p. 36), bounded by four similar planes meeting in two sorts of edges, but without solid angles. We may again distinguish three types of such prisms, according to which of the three axes the planes are parallel. Although they all belong equally to the prismatic type of forms, it is customary to call those whose planes are parallel to either of the lateral axes *domes*.

The prisms proper, whose planes are parallel to the

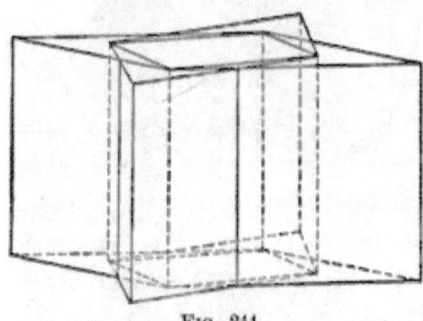

Fig. 244.

vertical axis, $\dot{c}$, are limiting forms, in one direction, of the pyramids, and must therefore be of three kinds, like these. The *unit prism*, whose lateral parameters are both unity, and whose symbols are $\breve{a} : \bar{b} : \infty \dot{c}$, ∞P, $\{110\}$; the *macroprisms*, whose lateral parameter n (>1) refers to the macrodiagonal axis, and whose symbols are $\breve{a} : n\bar{b} : \infty \dot{c}$, $\infty P\bar{n}$, $\{hk0\}$ ($h>k$); and the *brachyprisms*, whose lateral parameter n (>1) refers to the brachydiagonal axis, and whose symbols are $n\breve{a} : \bar{b} : \infty \dot{c}, \infty P\breve{n}, \{hk0\}$ ($h<k$) (Fig. 244).

The other two types of prismatic forms or domes are distinguished as *macrodomes* when their planes are parallel to the macrodiagonal axis, $\bar{b}$ (Fig 245); or as *brachydomes* when their planes are parallel to the brachydiagonal axis, $\breve{a}$ (Fig. 246). There is a vertical zone of each of these kinds of domes whose parameters m vary between zero and infinity. The general symbols of the macrodomes are $\breve{a} : \infty \bar{b} : m\dot{c}$, $mP\bar{\infty}$, $\{h0l\}$; and, those of the brachydomes, $\infty \breve{a} : \bar{b} : m\dot{c}$, $mP\breve{\infty}$,

{0kl}. The macro- and brachydomes together correspond in their positions to the tetragonal pyramids of the second order (p. 88).

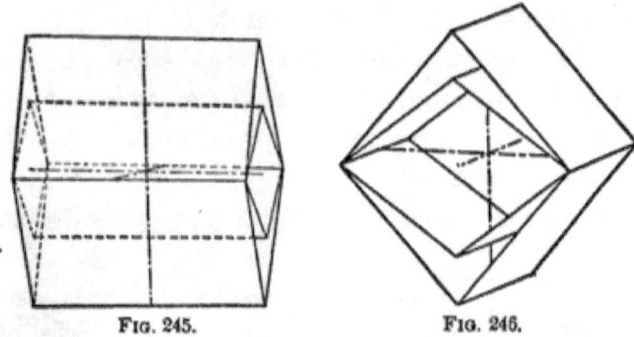

Fig. 245.   Fig. 246.

Pinacoids. These embrace the forms whose planes are simultaneously parallel to two axes, and whose parameter symbols must therefore contain two signs of infinity. Each of these is an open form bounded by but two planes which are parallel. The pinacoids of themselves therefore have neither edges nor solid angles. There are three kinds of orthorhombic pinacoids, as well as three kinds of pyramids and prisms. That one which is parallel to the vertical axis and macrodiagonal axis is called the *macropinacoid*. Its symbols are $\breve{a} : \infty \bar{b} : \infty \breve{c}, \infty P\bar{\infty}$, {100}. The form whose planes are parallel to the vertical and brachydiagonal axis is called the *brachypinacoid*, and its symbols are $\infty \breve{a} : \bar{b} : \infty \breve{c}, \infty P\bar{\infty}$, {010}. Finally, the planes parallel to both lateral axes are called *basal pinacoids*, and their symbols are $\infty \breve{a} : \infty \bar{b} : \breve{c} = \breve{a} : \bar{b} : 0\breve{c}$ (p. 29), OP, {001}. In their positions the orthorhombic pinacoids correspond to the faces of the isometric cube (p. 56); but the six planes must break up into three separate forms so soon as the axes cease to be equivalent and interchangeable.

The relations of limiting forms among the possible

types of orthorhombic holohedrons are exhibited in the following diagram:

Brachydiagonal Zone.	Brachypyramids.	Fundamental Zone.	Macropyramids.	Macrodiagonal Zone.
OP	OP	OP	OP	OP
$\frac{1}{m}P\breve{\infty}$ --	$\frac{1}{m}P\breve{n}$ --	$\frac{1}{m}P$ --	$\frac{1}{m}P\bar{n}$ --	$\frac{1}{m}P\bar{\infty}$
$P\breve{\infty}$ --	$P\breve{n}$ --	P --	$P\bar{n}$ --	$P\bar{\infty}$
$mP\breve{\infty}$ --	$mP\breve{n}$ --	mP --	$mP\bar{n}$ --	$mP\bar{\infty}$
$\infty P\breve{\infty}$ ---	$\infty P\breve{n}$ ---	∞P ---	$\infty P\bar{n}$ ---	$\infty P\bar{\infty}$

Prismatic Zone.

Orthorhombic Forms in Combination. As the grade of symmetry decreases from system to system, the number of planes in any crystal form also decreases; but at the same time the number of forms which occur in combination upon a single crystal is proportionately increased. If, however, the nature of the separate types of forms is clearly understood, there is no difficulty in deciphering these more complex combinations, both because of the greater simplicity of the forms themselves, and also on account of their zonal relation to each other.

The number of substances crystallizing in the orthorhombic system is very great. Of these we shall mention but a few, to serve as types of the more frequently occurring combinations.

The three orthorhombic pinacoids in combination would produce a figure identical with the cube in shape, and yet the crystallographic difference between the three pairs of planes may manifest itself by some physical peculiarity, as in the case of the three un-

THE ORTHORHOMBIC SYSTEM. 151

equal pinacoidal cleavages of anhydrite (calcium sulphate).

The fundamental prism, ∞P, {110}, is a common form on orthorhombic crystals. It may be combined with two pinacoids, as in Fig. 247, or with all three. Fig. 248 shows the fundamental prism (p) on a crystal

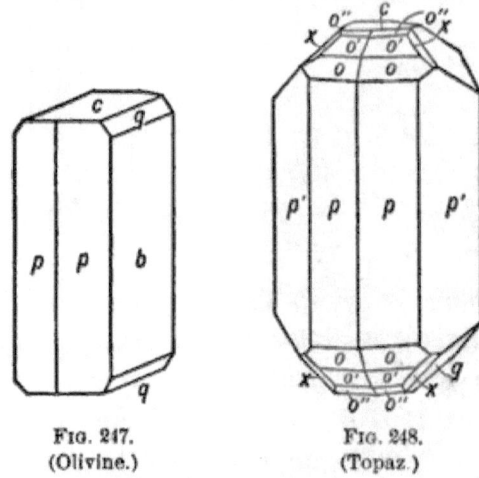

Fig. 247.
(Olivine.)

Fig. 248.
(Topaz.)

of topaz, in combination with another prism, $\infty P\breve{2}$, {120} (p'); the basal pinacoid, OP, {001} (c); the brachydome, $P\breve{\infty}$ {011} (g); and the four pyramids, P, {111} (o); $\frac{1}{2}P$, {112} (o'); $\frac{1}{3}P$, {113} (o''); and $\frac{2}{3}P\breve{2}$, {123} (x).

It not infrequently happens that the obtuse prismatic angle in the orthorhombic system approaches very nearly to 120°. Such a prism, combined with the brachypinacoid, would simulate a hexagonal prism. If these forms are terminated by fundamental pyramids and brachydomes, the whole combination often closely resembles a hexagonal crystal. Fig. 249 shows such a com-

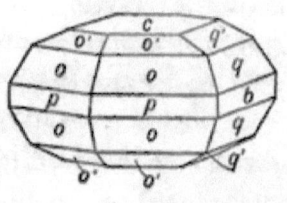

Fig. 249.
(Chalcocite.)

bination observed on the copper sulphide, chalcocite, with the forms: OP, $\{001\}$ (c); $\infty \bar{P} \infty$, $\{010\}$ (b); ∞P, $\{110\}$ (p); $2\bar{P}\infty$, $\{021\}$ (q); $\frac{2}{3}\bar{P}\infty$, $\{023\}$ (q'); P, $\{111\}$ (o); and $\frac{1}{3}P$, $\{113\}$ (o').

The orthorhombic combinations are far too numerous to be described systematically. A number are, however, appended, with the symbols of their planes, in order to familiarize the student with some of the more complex types.

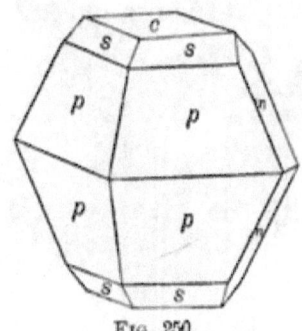

Fig. 250.

Fig. 250 (sulphur) shows the forms: $P, \{111\} (p)$; $\frac{1}{3}P, \{113\} (s)$; $OP, \{001\} (c)$; and $\bar{P}\infty, \{011\} (n)$.

Fig. 251 (silver sulphide, acanthite) shows a combination of the following forms: $\infty \bar{P}\infty$, $\{010\}$ (a); $\infty \bar{P}\infty$, $\{100\}$ (b); $OP, \{001\} (c)$; $\infty P, \{110\} (l)$; $\infty \bar{P}2$, $\{120\}$ (m); $\bar{P}\infty$, $\{010\}$ (o); $3\bar{P}\infty$, $\{031\}$ (e); $\frac{1}{2}\bar{P}\infty$, $\{102\}$ (d); $\bar{P}2$, $\{122\} (k)$; $P, \{111\} (p)$; $\frac{3}{2}P\frac{3}{2}, \{322\} (s)$; $\frac{1}{3}P, \{113\} (r)$; and $2\bar{P}4, \{142\} (n)$. If, however, as is still more common, the form m be assumed as the fundamental prism, instead of l, the axis running from back to front becomes twice as long and must then be regarded as the macrodiagonal instead of the brachydiagonal. In this case

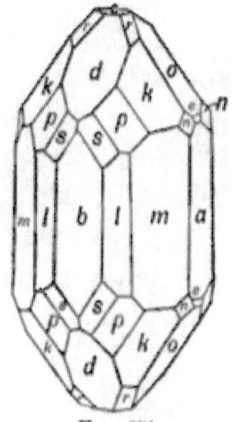

Fig. 251. (Acanthite.)

the above symbols for the planes of the figure become: $\infty \bar{P}\infty, \{100\} (a)$; $\infty \bar{P}\infty, \{010\} (b)$; $OP, \{001\} (c)$; $\infty \bar{P}2, \{120\} (l)$; $\infty P, \{110\} (m)$; $\bar{P}\infty, \{101\} (o)$; $3\bar{P}\infty, \{301\} (e)$; $\bar{P}\infty, \{011\} (d)$; $P, \{111\} (k)$; $2\bar{P}2, \{121\} (p)$; $3\bar{P}3, \{131\} (s)$; $\frac{2}{3}\bar{P}2, \{123\} (r)$; and $2\bar{P}2, \{211\} (n)$.

Fig. 252 reproduces a combination of orthorhombic forms observed on crystals of aluminium orthosilicate, andalusite (Al_2SiO_5): $\infty P\bar{\infty}$, {010} (a); $\infty P\bar{\infty}$, {100} (b); OP, {001} (c); ∞P, {110} (m); $\infty P\bar{2}$, {210} (l); $\infty P\check{2}$, {120} (n); $P\bar{\infty}$, {101} (r); P, {111} (p); $P\check{\infty}$, {011} (s); $2P\check{2}$, {121} (k).

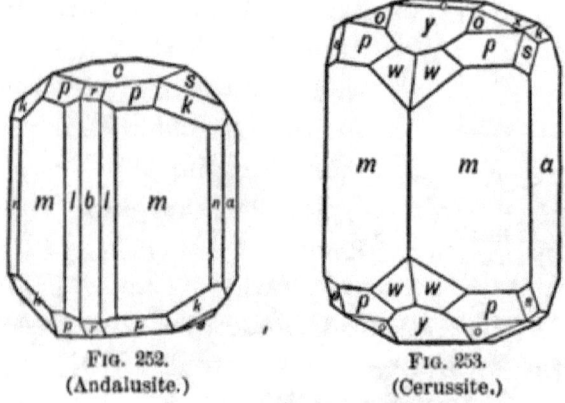

Fig. 252.
(Andalusite.)

Fig. 253.
(Cerussite.)

Fig. 253 shows a crystal of lead carbonate (cerussite) which is bounded by the following forms: OP, {001} (c); $\infty P\bar{\infty}$, {010} (a); ∞P, {110} (m); P, {111} (p); $\frac{1}{2}P$, {112} (o); $\frac{1}{2}P\bar{\infty}$, {102} (y); $\frac{1}{2}P\check{\infty}$, {012} (x); $P\check{\infty}$, {011} (k); $2P\check{2}$, {121} (s); $2P\bar{2}$, {211} (w).

Fig. 254 exhibits a combination occurring on the ferro-magnesian metasilicate, hypersthene [$(Fe,Mg)SiO_3$]. $\infty P\bar{\infty}$, {010} (a); $\infty P\bar{\infty}$, {100} (b); ∞P, {110} (m); $\infty P\check{2}$, {120} (n); $\frac{1}{4}P\check{\infty}$, {014} (h); P, {111} (o); $P\bar{2}$, {212} (c); $2P\bar{2}$, {241} (i); $\frac{3}{2}P\check{\frac{3}{2}}$, {232} (u).

Fig. 255 represents a crystal of the corresponding ferro-magnesian orthosilicate, olivine [$(Fe,Mg)SiO_4$]: $\infty P\bar{\infty}$, {010} (a); OP, {001} (c); ∞P, {110} (n); $\infty P\check{2}$, {120} (s); $\infty P\check{3}$, {130} (r); $P\bar{\infty}$, {101} (d); $P\check{\infty}$, {011} (h); $2P\check{\infty}$, {021} (k); $4P\check{\infty}$, {041} (i); P, {111} (e); $\frac{1}{2}P$, {112} (o); $2P\check{2}$, {121} (f); $3P\check{3}$, {131} (l).

Orthorhombic crystals are often disproportionately elongated in the direction of *one* axis, when they are

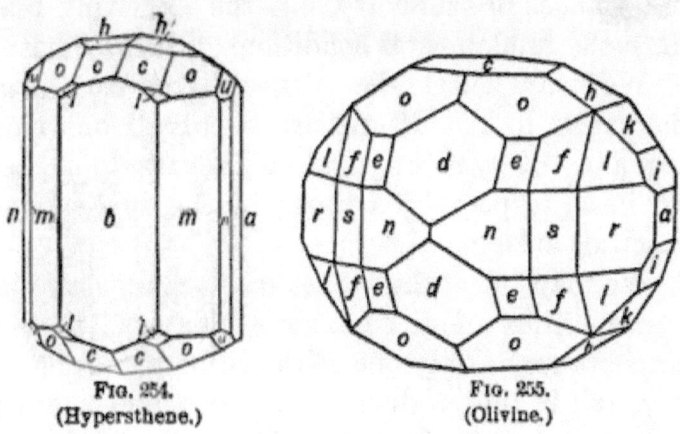

Fig. 254.
(Hypersthene.)

Fig. 255.
(Olivine.)

called *prismatic, columnar*, or *acicular* in their habit. In such cases the axis of elongation is generally selected as the vertical axis (see Fig. 251). In other cases the crystals are disproportionately shortened in the direction of one axis, when their habit is said to be *tabular* or *lamellar* (p. 14).

Hemihedrism in the Orthorhombic System.

Kinds of Hemihedrism. As the number of planes belonging to a single crystal form decreases with the symmetry of the system, the possibility of the existence of partial forms likewise decreases. Thus hemihedrism is of minor importance in the systems without any principal plane of symmetry, but it nevertheless exists.

It is possible to imagine one half of the eight planes which bound an orthorhombic pyramid to be chosen in three different ways.

THE ORTHORHOMBIC SYSTEM.

1. We may select the planes *alternately*. This produces, as we shall presently see, the disappearance of all the planes of symmetry, but the surviving planes satisfy the fundamental conditions of hemihedrism.

2. We may select the planes by *alternate pairs*. This results in the disappearance of all but one of the planes of symmetry, while the resultant forms differ in no respect from those of the succeeding or monoclinic system.

3. We may select the planes in *groups of four* about the extremities of either axis. This results in the disappearance of only one of the three planes of symmetry, and in the production of forms which are not hemihedral, but *hemimorphic*.

Sphenoidal Hemihedrism. There is therefore but a single kind of true hemihedrism possible in the orthorhombic system, and this is caused by the survival of alternate planes on the most general form, the orthorhombic pyramid. The extension of these planes until they intersect must result in the production of an asymmetric figure, bounded by four similar scalene triangles, meeting in three sorts of edges. Such a form is exactly analogous to the isometric tetrahedron or tetragonal sphenoid (p. 101), and is called the *orthorhombic sphenoid*. The two sphenoids derivable in this

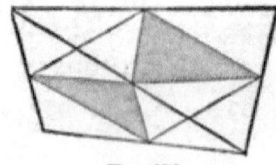
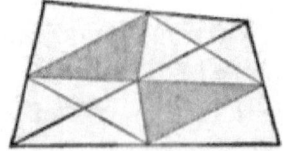

Fig. 256.   Fig. 257.

way from every orthorhombic pyramid are, on account of their lack of all planes of symmetry, enantiomorphous forms (p. 66) (Figs. 256 and 257). Their general parameter and index symbols are:

$$\frac{m\breve{P\bar{n}}}{2}r, \quad \kappa\{hkl\}; \quad \text{and} \quad \frac{m\breve{P\bar{n}}}{2}l, \quad \kappa\{h\bar{k}l\}.$$

It is evident that this method of selection can produce no new forms from orthorhombic prisms or domes, since each plane of these forms corresponds to two contiguous pyramidal planes. Still less can it give rise to new forms when applied to the pinacoids, whose planes correspond to four pyramidal faces.

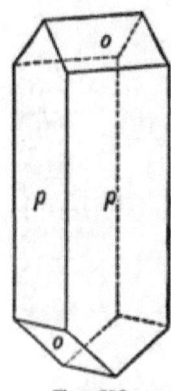

Fig. 258.

Orthorhombic substances which exhibit sphenoidal hemihedrism are not altogether uncommon, especially among organic salts. As examples of this method of crystallization may be mentioned the silicate, leucophane; the hydrous sulphates of magnesium (Epsom salts), $MgSO_4 + 7$ aq (Fig. 258), and zinc (zinc vitriol), $ZnSO_4 + 7$ aq; acid potassium tartrate (cream of tartar); potassium sodium tartrate (Rochelle salts); tartar emetic; lactose; and mycose.

Sulphur crystals also rarely exhibit a sphenoidal development of their planes.

Hemimorphism. Hemimorphism in the direction of one of the three orthorhombic axes is also not uncommon. In such cases it is usual to select the hemimorphic axis as the vertical axis. The two planes of symmetry which intersect in this axis remain in hemimorphic crystals as such, while the plane normal to it ceases to be a plane of symmetry.

THE ORTHORHOMBIC SYSTEM. 157

As examples of orthorhombic hemimorphism may be cited ammonium-magnesium phosphate (struvite), $NH_4MgPO_4 + 6$ aq (Fig. 259), which shows above (antilogue pole) the faces: $P\bar{\infty}$, $\{101\}$ (r); $P\check{\infty}$, $\{011\}$ (q); $4 P\check{\infty}$, $\{041\}$ (q'); while below (analogue pole) are the forms $OP, \{001\}$ (c); $\frac{1}{3}P\bar{\infty}, \{103\}$ (r'); and $\infty P\check{\infty}$, $\{010\}(b)$.

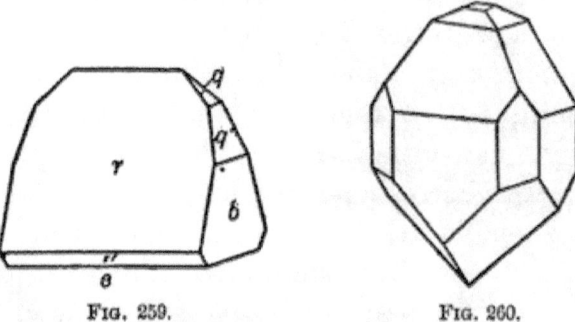

FIG. 259.   FIG. 260.

Other hemimorphic substances are basic zinc silicate (calamine, hemimorphite), $Zn_2(OH)_2SiO_3$ (Fig. 260); resorcine, $C_6H_6O_2$; triphenylmethan, $(C_6H_5)CH$; lactose, $C_{12}H_{22}O_{11}$. The last-named substance is both hemihedral and hemimorphic.

CHAPTER VII.

THE MONOCLINIC SYSTEM.*

Symmetry and Axes. No complete crystal form, i.e. one bounded by pairs of parallel planes (p. 18), can possess two planes of symmetry, without at the same time possessing three. The grade of symmetry next lower than the orthorhombic, which can possibly characterize a crystal system, must therefore be produced by a *single* plane of symmetry.

The existence of such a single plane of symmetry determines the direction normal to it as an axis of symmetry, and this is the only direction which is fixed for the whole system (Fig. 261). The other two necessary axes of reference must lie in the plane of symmetry, but their directions in this plane are a matter of arbitrary choice, to be decided in the case of each substance as shall be most convenient. Both of these other axes must therefore, in all cases, be perpendicular to the axis of symmetry, but they may make any angle whatever with each other.

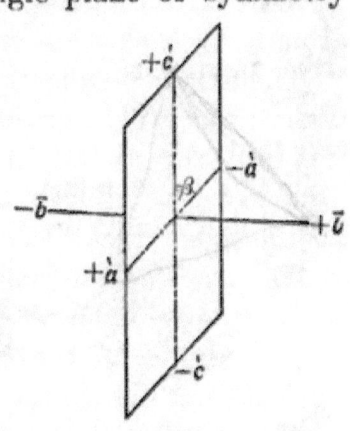

FIG. 261.

It is customary to place crystals having only one plane of symmetry in such a position that this plane

---

* Known also as the *monosymmetric, clinorhombic* or *oblique* system.

shall stand vertical, and the axis normal to it shall run from right to left. Furthermore, one of the two directions which are chosen as axes in the plane of symmetry is made vertical, and the other is so placed that it inclines downward toward the front (Fig. 261). In this way we see that the three axes directly deducible from the single plane of symmetry are two directions at right angles, and a third which is at right angles to the first but inclined at any angle to the second. Hence crystal forms whose planes are referable to such axes are called *monoclinic*.

The axis of symmetry, which is always made to run from right to left, is designated by the letter b surmounted by a straight line, $\bar{}$, and called the *orthodiagonal axis*. The axis which is made vertical is represented by the letter $\dot{c}$, as in the orthorhombic system, and called the *vertical axis*, while the oblique axis is designated by a surmounted by an oblique sign, $\grave{}$, and called the *clinodiagonal axis*. This position differs from that in the orthorhombic system in so far as the axis $\bar{b}$ may or may not be longer than $\dot{a}$.

Fundamental Form and Crystallographic Constants. Just as in the orthorhombic system, so in the monoclinic, *any* pyramid occurring on a crystal *may* be assumed as a ground-form, from whose angles the axial ratio for the particular substance may be calculated. This ground-form here determines not merely the relative unit lengths of the three axes $\dot{a} : \bar{b} (=1) : \dot{c}$, but also the inclination of

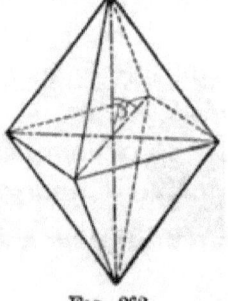

Fig. 262.

the two axes $\dot{a}$ and $\dot{c}$ to each other (Fig. 262). This angle, which is designated by the Greek letter β, is a crystallographic constant for each monoclinic crystal series, just as are the two irrational quotients $\frac{a}{b}$ and $\frac{c}{b}$.*

Although any monoclinic pyramid could possibly serve as a fundamental form, the choice of this is not altogether arbitrary, since it is very desirable that the axes $\dot{a}$ and $\dot{c}$ should be parallel to prominent planes, which then become pinacoids or prisms. One of them, at least, is usually determined by the habit of the crystal, and is made to coincide with a direction of elongation or cleavage. If different authors select different ground-forms for crystals of the same substance, the crystallographic constants derived from these must always bear a rational relation to one another.

Derivation of the Holohedral Monoclinic Forms. The most general monoclinic symbol, $n\dot{a} : \bar{b} : m\dot{c}$, stands for only half as many planes as are represented by the corresponding orthorhombic symbol (p. 145). As was explained in Chapter II (p. 35), the presence of any plane, A, making any angle with a single plane of symmetry, $WXYZ$, necessitates the presence of the similar plane B; while to both A and B there must be a parallel plane A' and B' (Fig. 263). Four is therefore the greatest number of planes that can belong to any monoclinic form.

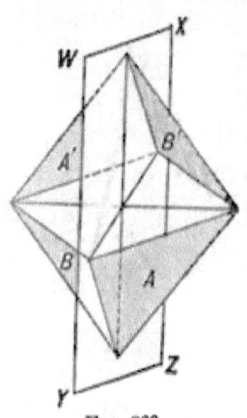

Fig. 263.

---

* The process of calculating the crystallographic constants from the angles of the fundamental form becomes, in the monoclinic and triclinic systems, too complex to be given here, and must be sought in works like those of Dana, Klein or Liebisch, where mathematical crystallography is treated of at length.

The four planes composing any monoclinic pyramid belong either altogether to the acute or altogether to the obtuse octants. Such forms cannot of themselves enclose space; but if their planes are extended, they form open prisms (Fig. 264); and in fact they would really be monoclinic prisms if either of the axes lying in the plane of symmetry had been selected so as to run parallel to their intersection edges.

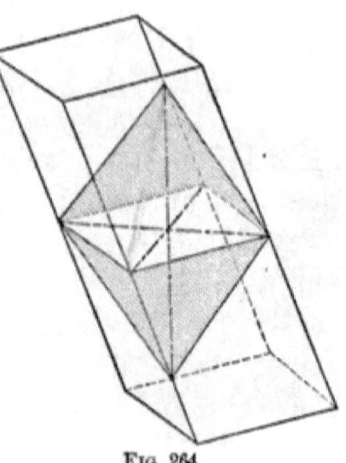

Fig. 264.

Monoclinic pyramids, because they occupy but half of the eight octants, are called *hemi-pyramids*, and are furthermore distinguished as *positive* and *negative*, according as their planes belong to the *acute* or *obtuse* octants.* The parameter and index symbols of corresponding positive and negative hemi-pyramids are $+ mPn$, $\{\bar{h}kl\}$ and $- mPn$, $\{hkl\}$. Two such forms together correspond to the planes of the orthorhombic pyramid, but they are nevertheless completely independent and holohedral. Their planes are of different shapes, and either alone entirely satisfies the conditions of monoclinic symmetry.

The possible monoclinic holohedrons, derived from the most general symbol by giving limiting values to

---

* This universally accepted nomenclature is unfortunate, since the four-faced monoclinic pyramids are in no sense hemihedral. The upper positive form likewise cuts the negative end of the clinodiagonal axis, but this usage was adopted by Naumann so as to make the cos β (when $< 90°$) a positive quantity, since this is an important factor in the calculation of monoclinic forms.

one or both of its variable parameters, agree exactly with those in the orthorhombic system, except that all forms whose planes belong exclusively to acute or obtuse octants become positive and negative hemi-forms. All forms, on the other hand, whose planes lie simultaneously in acute and obtuse octants are identical in the two systems.

Pyramids. There are three sorts of positive and negative monoclinic hemi-pyramids corresponding to the three sorts of orthorhombic pyramids; and, like them, named from the lateral axis to which their larger parameter, n, refers. These are:

1. The zone or series of unit pyramids, both of whose lateral parameters are unity. Their general symbols are $+mP$, $\{\bar{h}hl\}$ and $-mP$, $\{hhl\}$. They extend, according to the value of the vertical parameter, m, between the unit prism where m becomes ∞ and the basal pinacoid where it becomes 0.

2. On one side of the unit pyramids lie those whose larger lateral parameter, n, refers to the ortho-diagonal axis. These are called *hemi-orthopyramids*, and their general symbols are $+mP\bar{n}$, $\{\bar{h}kl\}$ ($h>k$) and $-mP\bar{n}$, $\{hkl\}$ ($h>k$).

3. On the opposite side of the unit pyramids are such as have their larger lateral parameter n referring to the clino-diagonal axis. These are called *hemi-clinopyramids*, and their symbols are $+mP\bar{n}$, $\{\bar{h}kl\}$ ($h<k$) and $-mP\bar{n}$, $\{hkl\}$ ($h<k$).

Prisms. There are three sorts of prismatic types whose planes are parallel to one of the three axes of reference. As in the orthorhombic system, those which are parallel to the vertical axis are called *prisms* proper, and the others *domes*.

THE MONOCLINIC SYSTEM.

The three kinds of vertical prisms correspond to the three kinds of monoclinic pyramids, viz.: *unit prisms*, whose symbols are $\dot{a} : \bar{b} : \infty \dot{c}, \infty P, \{110\}$; *orthoprisms*, whose symbols are $\dot{a} : n\bar{b} : \infty \dot{c}, \infty P\bar{n}, \{hk0\}$ $(h > k)$; and *clinoprisms*, whose symbols are $n\dot{a} : \bar{b} : \infty \dot{c}, \infty P\bar{n}, \{hk0\}$ $(h < k)$. The planes of all of these forms belong at once to both an acute and an obtuse octant and hence do not in any way differ from the four-sided orthorhombic prisms.

The monoclinic domes are named from the axes to which their planes are parallel, *orthodomes*, whose symbols are $\dot{a} : \infty \bar{b} : m\dot{c}, mP\bar{\infty}, \{h0l\}$; and *clinodomes*, whose symbols are $\infty \dot{a} : \bar{b} : m\dot{c}, mP\check{\infty} \{0kl\}$. The orthodomes have their planes confined entirely to two acute or to two obtuse octants, and hence consist of two independent forms of two planes each, $+ mP\bar{\infty}, \{\bar{h}0l\}$ and $- mP\bar{\infty} \{h0l\}$, called positive and negative hemi-orthodomes. The clinodomes, on the other hand, have their planes lying at once in an acute and an obtuse octant, and cannot therefore break up into hemi-forms.

Pinacoids. These, as in the other systems, consist of pairs of planes which are parallel to two of the axes. Their planes must belong simultaneously to two acute and two obtuse octants, so that they cannot differ from the corresponding orthorhombic forms (p. 149). They are known as the *orthopinacoid*, parallel to the vertical and orthodiagonal axis $\dot{a} : \infty \bar{b} : \infty \dot{c}, \infty P\bar{\infty}, \{100\}$; the *clinopinacoid*, parallel to the vertical and clinodiagonal axis $\infty \dot{a} : \bar{b} : \infty \dot{c}, \infty P\check{\infty}, \{010\}$; and the *basal pinacoid*, parallel to the axis of symmetry (the orthodiagonal) and that direction which has been assumed as the clinodiagonal axis, $\infty \dot{a} : \infty \bar{b} : \dot{c}$, or $\dot{a} : \bar{b} : 0\dot{c}, OP, \{001\}$.

The relations of limiting forms in the monoclinic system are shown in the following diagram:

Clinodiagonal Zone.	Clinopyramids.	Fundamental Zone.	Orthopyramids.	Orthodiagonal Zone.
OP	OP	OP	OP	OP
$\frac{1}{m}P\infty$	$\pm\frac{1}{m}P\bar{n}$	$\pm\frac{1}{m}P$	$\pm\frac{1}{m}P\bar{n}$	$\pm\frac{1}{m}P\bar{\infty}$
$P\infty$	$\pm P\bar{n}$	$\pm P$	$\pm P\bar{n}$	$\pm P\bar{\infty}$
$mP\infty$	$\pm mP\bar{n}$	$\pm mP$	$\pm mP\bar{n}$	$\pm mP\bar{\infty}$
$\infty P\infty$	$\infty P\bar{n}$	∞P	$\infty P\bar{n}$	$\infty P\bar{\infty}$

Prismatic Zone.

Monoclinic Forms in Combination. Monoclinic combinations are more varied, and at the same time more common, than those of any other system. It is, however, impossible to convey a satisfactory idea of their proportions by means of plane figures; and hence the use of models and actual crystals is more necessary than ever to render them completely intelligible. The first step in deciphering the combinations presented by such models or crystals is to bring them into correct position, i.e. to seek out their plane of symmetry and to place it vertically and at the same time to direct it toward the observer. The orthodiagonal axis will then run horizontally and from right to left. Thus far the position is imperative for all monoclinic crystals, but the selection of the remaining axes is a matter of convenience in each particular case. Of all the planes which are perpendicular to the plane of symmetry (that is, which belong to the orthodiagonal zone) *any* parallel pair may be chosen as the orthopinacoids, and any other pair as the basal pinacoids. This choice will of course be regulated by the habit of the crystal, and, in the case of all common substances, it is already established by a usage which cannot be ignored. Whichever planes are made the orthopinacoids will determine the direction of the vertical axis, while those selected as the basal pinacoids must be placed so as to slope downward toward the observer, thus conditioning the direction of the clinodiagonal axis. When the crystal has in this manner been brought completely into position, it is possible to designate its other planes. All those lying in the orthodiagonal zone, except the pinacoids, become positive or negative hemi-orthodomes. All other planes really belong to one class of four-sided prismatic forms, and which of them become prisms, which clinodomes, and which pyramids, depends entirely upon the choice of the ortho- and basal pinacoids.

On the crystal of iron-vitriol represented in Fig. 265 the only plane whose value is absolutely fixed is the plane of symmetry, or clinopinacoid (b). It is customary to make c the basal pinacoid, and p the fundamental prism, whence q becomes a clinodome, o a negative hemi-pyramid, and r' and r plus and minus hemi-orthodomes. We might, however, turn the crystal so as to make c the orthopinacoid, q the prism, and p a clinodome;

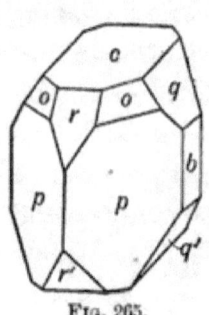

Fig. 265.

or we might even make r' the basal pinacoid, and r the orthopinacoid, when o would become the prism, c a hemi-orthodome, and p and q both pyramids.

The following examples of monoclinic combinations observed on natural crystals will be found useful.

Fig. 266 shows a crystal of orthoclase with the forms $\infty P \bar{\infty}$, $\{010\}$ (M); OP, $\{001\}$ (P); ∞P, $\{110\}$ (T); $+ P\bar{\infty}$, $\{10\bar{1}\}$ (x); $+ 2P\bar{\infty}$, $\{20\bar{1}\}$ (y); $+ P$, $\{11\bar{1}\}$ (o); $2P\bar{\infty}$, $\{021\}$ (n).

Fig. 267 represents a crystal of arsenic disulphide (realgar), As_2S_2, with the forms OP, $\{001\}$ (P);

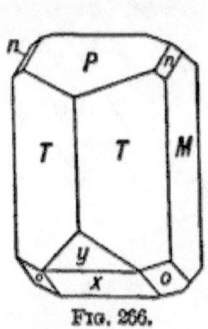

Fig. 266.

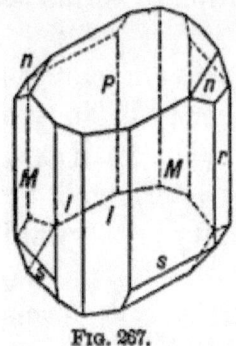

Fig. 267.

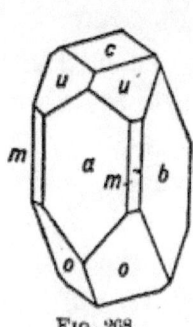

Fig. 268.

$\infty P\bar{\infty}$, $\{010\}$ (r); ∞P, $\{110\}$ (M); $\infty P\bar{2}$, $\{210\}$ (l); $P\bar{\infty}$, $\{011\}$ (n); and $+P$, $\{11\bar{1}\}$ (s).

Fig. 268 represents a crystal of calcium-magnesium metasilicate (diopside), $(CaMg)SiO_3$, the symbols of whose forms are: OP, $\{001\}$ (c); $\infty P\bar{\infty}$, $\{010\}$ (b); $\infty P\bar{\infty}$, $\{100\}$ (a); ∞P, $\{110\}$ (m); $+2P$, $\{\bar{2}21\}$ (o); and $-P$ $\{111\}$ (u).

Fig. 269 shows a crystal of calcium silico-titanate (sphene), $CaTiSiO_5$, whose forms are: OP, $\{001\}$ (y); $\infty P\bar{\infty}$, $\{010\}$ (g); $\infty P\bar{\infty}$, $\{100\}$ (c); ∞P, $\{110\}$ (r); $P\bar{\infty}$, $\{011\}$ (e); $+P$, $\{11\bar{1}\}$ (t); $-P$, $\{111\}$ (n); $-2P$, $\{221\}$ (η).

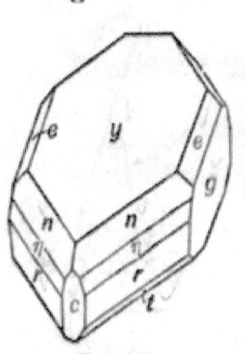

Fig. 269.

Monoclinic crystals are very often elongated parallel to some direction in the plane of symmetry, which direction is then generally made the vertical axis (see Fig. 266). In cases where the basal pinacoid is fixed by some physical property like cleavage, the elongation is in the direction of the clinodiagonal axis. This is often the case with crystals of the mineral feldspar, as is shown in Fig. 270 which represents nearly the same combination of forms as Fig. 266.

In still other cases the elongation of the crystal may be in the direction of the orthodiagonal axis, whose position is determined by the plane of symmetry. A crystal of the silicate mineral, epidote, shows this (Fig. 271). Its forms are: OP, $\{001\}$ (M); $\infty P\bar{\infty}$, $\{100\}$ (T); $+2P\bar{\infty}$, $\{20\bar{1}\}$ (l); $+P\bar{\infty}$, $\{10\bar{1}\}$ (r); ∞P, $\{110\}$ (z); $+P$, $\{\bar{1}11\}$ (n).

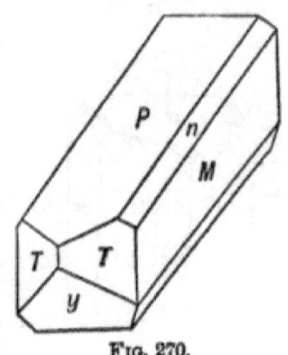

Fig. 270.

Monoclinic crystals may also be flattened in the direction of their axis of symmetry, as in the case of sanidine; or in a direction perpendicular to it, as in the case of mica.

Fig. 271.

HEMIHEDRISM IN THE MONOCLINIC SYSTEM.

Kinds of Hemihedrism. It is possible to select one half of the four planes composing a monoclinic hemipyramid in three and only three different ways, which are illustrated in the three following figures (272, 273 and 274).

1. We may select the two parallel planes B and B' (Fig. 263) belonging to two diagonally opposite octants as shown in Fig. 272, which would produce forms in no way differing from those in the following or triclinic system.

2. We may select the two planes A and B at one extremity of the vertical axis; or, in other words, the

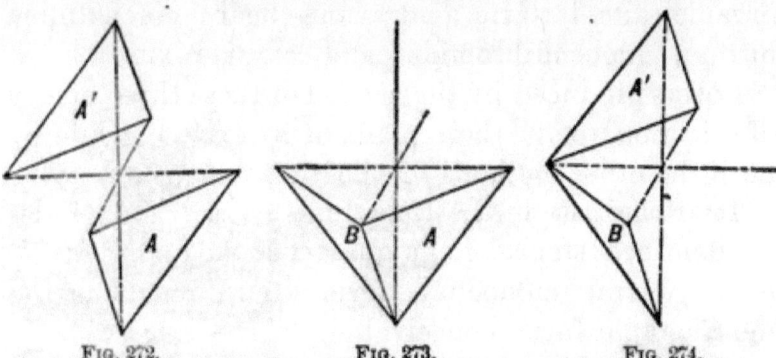

Fig. 272.   Fig. 273.   Fig. 274.

two which intersect in the plane of symmetry (Fig. 273). By such a selection the surviving planes are still equally distributed about both extremities of the axis of symmetry, and they therefore satisfy all the conditions of a true hemihedrism (p. 41). Fig. 275 shows such a hemihedral crystal of pyroxene, whose forms are the same as those represented in Fig. 268.

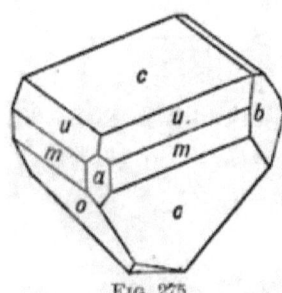

Fig. 275.

3. We may select the two planes B and A', which intersect at one end of the axis of symmetry (Fig. 274). This will result in the production, not of hemihedrism, but of hemimorphism in the direction of the axis of symmetry. Such a development is illustrated by pentacid alcohol (quercite), $C_6H_{12}O_5$. The forms on the crystal of this substance shown in Fig. 276 are: OP, $\{001\}$ (c); ∞P, $\{110\}$ (p); $+ P\check{\infty}$, $\{10\bar{1}\}$ (r); and $P\grave{\infty}$, $\{011\}$ (q). The latter form occurs only on

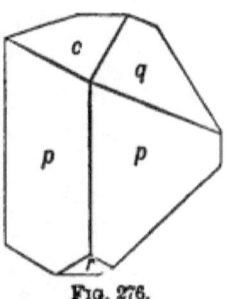

Fig. 276.

one side of the plane of symmetry. Other examples of monoclinic hemimorphism are afforded by the organic salts, tartaric acid; cane-sugar; cinchotinine nitrate; cinchendibromide; and campheroxim.

Forms produced by the second of these three modes of selection retain their plane of symmetry, while by both the other methods of selection this disappears.

Tetartohedrism in the Monoclinic System. The independent occurrence of one quarter of the planes of the most general monoclinic form would result in the most simple form conceivable, i.e. a single plane. This would not be separable from hemihedrism in the triclinic system, where each holohedron consists of two parallel planes. The repeated occurrence of isolated pyramidal planes on crystals of cane-sugar has led some authors to consider this substance as possibly an example of such crystallization (Fig. 277).

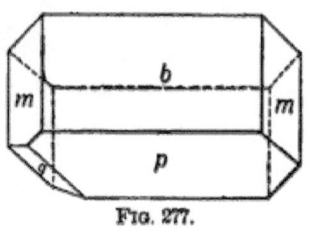

Fig. 277.

CHAPTER VIII.

THE TRICLINIC SYSTEM.*

Symmetry and Axes. The only possible grade of symmetry lower than that possessed by the monoclinic system (p. 158) consists in the absence of any plane of symmetry whatever. We have already encountered in all the preceding systems *partial* forms which were altogether without symmetry; but such forms do not generally possess the property, which belongs to every crystallographic holohedron, of having their planes distributed in parallel pairs (p. 18). The forms of the present system, if they are to be regarded as holohedral, must possess this property, as they do.

Since there is no plane of symmetry, the presence of any face necessitates only the presence of its opposite parallel face (see Fig. 31, p. 35); so that all forms of the triclinic system are alike in consisting of only two parallel planes.

The selection of the three axes of reference is altogether a matter of convenience, since none of them can be an axis of symmetry. Any three pairs of planes occurring on a crystal may be chosen as the pinacoids, and so be made to determine the positions of the axes. The axes will therefore differ, not merely in their unit length, but also in their directions for all triclinic crystals of different substances (crystal series); while their

---

* Also called the *asymmetric, clinorhomboidal* or *anorthic* system.

directions may also differ for crystals of the same substance, if they are differently chosen by different authors. But however the axes are selected, they must always be of different unit lengths, and must also be obliquely inclined to each other, whence the designation of this system as the *triclinic*. When their directions are decided upon, it is customary to place the crystal in such a position that one axis shall stand vertically (the *vertical axis*, c) (Fig. 278); while the longer of the two remaining axes inclines downward toward the right (*macrodiagonal*, $\bar{b}$), and the shorter downward toward the front (*brachydiagonal*, $\breve{a}$). The three oblique interaxial angles are designated by Greek letters as follows: $c \wedge \bar{b} = \alpha$; $c \wedge \breve{a} = \beta$; $\bar{b} \wedge \breve{a} = \gamma$. When in the above position, all the interaxial angles in the upper right front octant will be greater than 90°.

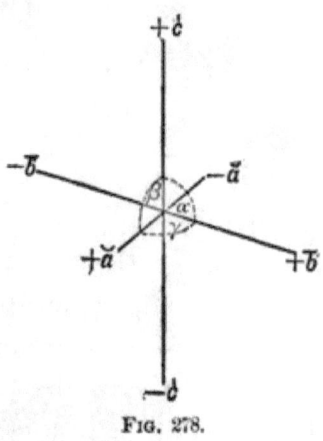

FIG. 278.

Fundamental Form and Crystallographic Constants. In the triclinic system there are five constants to be determined for each crystal series. These are the three interaxial angles, α, β and γ, whose values are given by the inclinations of the faces selected as pinacoids, and the two irrational quotients, $\frac{a}{b}$ and $\frac{c}{b}$, which determine the axial ratio, $\breve{a} : \bar{b} : c$, just as they do in the orthorhombic and monoclinic systems. These five constants cannot all be determined from a single form; nor is this necessary, since a triclinic form cannot occur except in combination with others. If the directions of the axes are known, both ratios may be calculated from any pyramidal face which is selected as the unit pyramid; or the first may be more easily obtained from a prism, and the second from a dome face. The macrodiagonal axis, $\bar{b}$, is here, as in the preceding systems, assumed as the unity for the axial ratio.

Derivation of the Holohedral Triclinic Forms. When the fundamental forms have been selected for any triclinic substance, and its crystallographic constants thereby determined, the designation of all other possible forms agrees closely with that employed in the orthorhombic system. We have but to remember that no triclinic crystal form can consist of more than two planes, and that these must always be parallel, as explained on p. 35.

Pyramids. Any plane on a triclinic crystal which intersects all three of the directions assumed as the crystallographic axes is a pyramid. Furthermore, any pyramid may be selected as the fundamental form, as was stated above. When this selection has been made, there will be a possible zone of unit pyramids whose lateral parameters are the same, but whose vertical parameters are different from those of the fundamental form.

Pyramidal planes occurring on one side of the unit pyramids, with their parameters on the brachydiagonal axis greater than unity, are called *brachypyramids;* while those occurring on the other side of the unit pyramids, with their parameters on the macrodiagonal axis greater than unity, are called *macropyramids*, as in the orthorhombic system (p. 147).

Since the entire absence of symmetry in the triclinic system necessitates but two parallel planes for any complete holohedral form, no triclinic pyramid can occupy more than two of the eight octants into which the axial planes divide space. By the oblique intersections of these axial planes, however, these octants are not all similar, as in the orthorhombic system, but fall into four pairs of dissimilar octants, the similar members of each pair being diagonally opposite. The two parallel pyramidal planes occurring in any one of the four pairs of octants are entirely independent of any planes occurring in the other octants, even when these have the same parameters. Because the complete triclinic pyramid can occupy but one quarter of the octants it is called, by analogy with the monoclinic hemi-pyramid (p. 161), a *tetra-pyramid* (see Fig. 31, p. 35).

The symbols for the triclinic pyramids are like those of the other systems, except that an accent is written near Naumann's initial P, to designate which one of the four dissimilar octants on the front of the crystal the form belongs to. Thus the four triclinic tetra-pyramids whose parameters are all unity are written:

$P'\{111\}$ for the upper right-hand octant;
$'P\{1\bar{1}1\}$ for the upper left-hand octant;
$P_{,}\{11\bar{1}\}$ for the lower right-hand octant;
$_{,}P\{1\bar{1}\bar{1}\}$ for the lower left-hand octant.

Prisms. All forms of the prismatic type (p. 36) are composed of planes which are parallel to one of the crystallographic axes, and therefore common to two contiguous octants. In the orthorhombic system these forms are all composed of four planes which correspond, in the triclinic system, to two independent forms of two planes each. By analogy with the orthorhombic designation (p. 148), these are called *hemi-prisms* when they are parallel to the vertical axis; *hemi-brachydomes* when they are parallel to the shorter lateral axis; and *hemi-macrodomes* when they are parallel to the longer lateral axis. Like the pyramids, their position is indicated by accents written in the symbols, though two of these are now necessary to show the two octants to which their planes belong. Thus the symbols for the prismatic forms two of whose parameters are unity become in the triclinic system

$\infty P,'$ {110} right-hand hemi-prism;
$,\infty P$ {1$\bar{1}$0} left-hand hemi-prism;
$,P\breve{\infty}'$ {011} right-hand upper hemi-brachydome;
$'P\breve{\infty},$ {0$\bar{1}$1} left-hand upper hemi-brachydome;
$'P\bar{\infty}'$ {101} upper front hemi-macrodome;
$,P\bar{\infty},$ {10$\bar{1}$} lower front hemi-macrodome.

Pinacoids. The pinacoidal planes are parallel to two axes, and therefore belong equally to four contiguous octants. The pinacoids consist of but two parallel planes in the orthorhombic system (p. 149), and therefore cannot be different, except in their relative inclinations, in the triclinic system. Their names and symbols are identical with those in the orthorhombic system, viz.:

$\infty P\breve{\infty}$ {010} parallel to $\breve{a}$ and $\dot{c}$, the brachy-pinacoid;
$\infty P\bar{\infty}$ {100} parallel to $\bar{b}$ and $\dot{c}$, the macro-pinacoid;
OP {001} parallel to $\breve{a}$ and $\bar{b}$, the basal pinacoid.

In the following diagram the relation of the different form-types possible in a triclinic crystal series is represented as it has been in each of the preceding systems.

THE TRICLINIC SYSTEM.

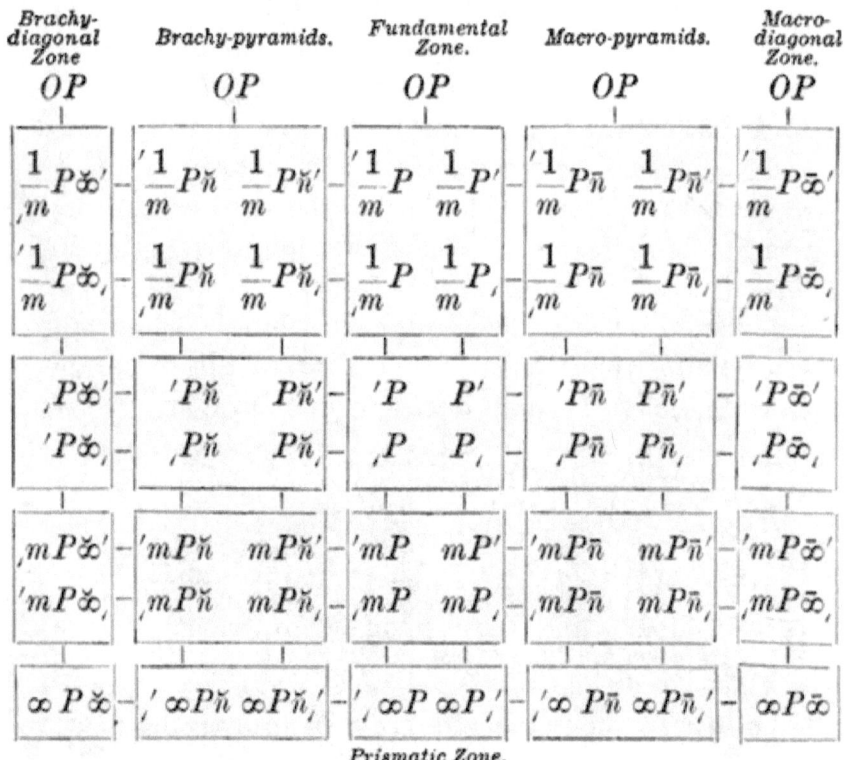

The derivation of the index symbols, corresponding to these parameter symbols of Naumann, is in all respects the same as in the preceding systems, and can therefore present no difficulty.

Triclinic Forms in Combination. Since the number of planes belonging to each form is in the triclinic system the least possible, the variety of possible combinations is proportionately greater. Such combinations are not, however, difficult to decipher. Familiarity with them can only be acquired by study of figures, models, or crystals of triclinic substances, a few examples of which are here appended by way of illustration.

Fig. 279 shows a combination observed on calcium hyposulphite ($CaS_2O_3 + 6$ aq). If we assume the face

c as the basal pinacoid OP {001}, and b as the brachypinacoid, $\infty P\breve{\infty}$ {010}, then the symbols of the other planes become: $\infty P_{\prime}'$, {110} (p); $_{\prime}'\infty P$, {1$\bar{1}$0} (p'); and $_{\prime}P\breve{\infty}'$, {011} ($q$).

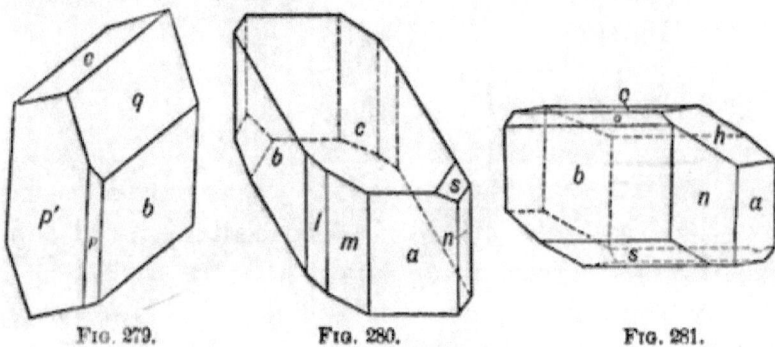

Fig. 279.   Fig. 280.   Fig. 281.

Fig. 280 reproduces a crystal of hydrous copper sulphate whose forms are: OP, {001} (c); $\infty P\bar{\infty}$, {100} (a); $\infty P\breve{\infty}$, {010} (b); $\infty P_{\prime}'$, {110} (n); $_{\prime}'\infty P$, {1$\bar{1}$0} (m); $_{\prime}'\infty P\breve{2}$, {1$\bar{2}$0} ($l$); and P', {111} (s).

Fig. 281 shows a crystal of the manganese metasilicate (rhodonite), $MnSiO_3$, whose forms are: OP, {001} (c); $\infty P\bar{\infty}$, {100} (b); $\infty P\breve{\infty}$, {100} (a); $\infty P_{\prime}'$, {110} (n); $_{\prime}P\breve{\infty}'$, {011} ($h$); $'P\bar{\infty}'$, {101} (o); and $_{\prime}P\bar{\infty}_{\prime}$, {10$\bar{1}$} ($s$).

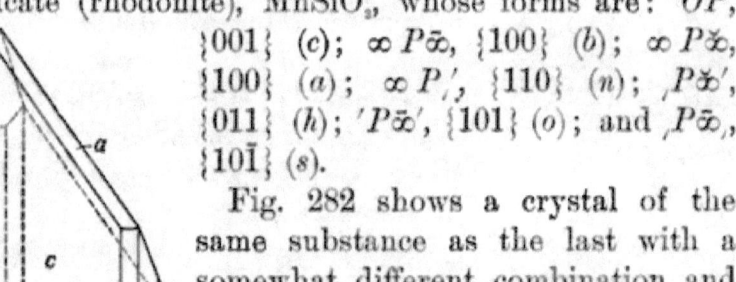

Fig. 282.

Fig. 282 shows a crystal of the same substance as the last with a somewhat different combination and placed in a different position. The same planes have the same letters as in Fig. 281, but they now receive different symbols, as follows: OP, {001} (a); $\infty P\bar{\infty}$, {100} (o); $\infty P\breve{\infty}$, {010} (s); $\infty P_{\prime}'$, {110} (b); $_{\prime}'\infty P$, {1$\bar{1}$0} (c); $2P_{\prime}$, {22$\bar{1}$} (n); and $_{\prime}2P$, {2$\bar{2}\bar{1}$}.

Fig. 283 shows a complicated crystal of calcium aluminium unisilicate (anorthite), $CaAl_2Si_2O_8$, whose planes are given the symbols: OP, $\{001\}$ (P); $\infty P\breve{\infty}$, $\{010\}$ (M); $\infty P_{,\prime}'$, $\{110\}$ (l); $,'\infty P$, $\{1\bar{1}0\}$ (T); $\infty P\breve{3}'$, $\{130\}$ (f); $,'\infty P\breve{3}$, $\{1\bar{3}0\}$ (z); P', $\{111\}$ (m); $'P$, $\{1\bar{1}1\}$ (a); $P_{,\prime}$, $\{11\bar{1}\}$ (o); $_{,\prime}P$, $\{1\bar{1}\bar{1}\}$ (p); $4P\breve{2}_{,\prime}$, $\{24\bar{1}\}$ (v); $'P\breve{\infty}'$, $\{101\}$ (t); $_{,\prime}2P\breve{\infty}_{,\prime}$, $\{20\bar{1}\}$ (y); $_{,\prime}P\breve{\infty}'$, $\{011\}$ (e); $_{,\prime}2P\breve{\infty}'$, $\{021\}$ (r); and $'P\breve{\infty}_{,\prime}$, $\{0\bar{1}1\}$ (n).

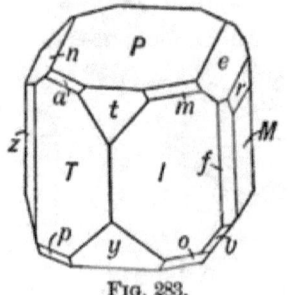

Fig. 283.

The obliquity of the axes in this crystal is so slight that it approaches closely to its limiting form in the monoclinic system, which actually occurs on the corresponding potassium feldspar, orthoclase.

How crystals may find their limiting forms in systems of a higher grade of symmetry has already been explained in Chap. IV (p. 91). To make this clearer, the student will find it a useful practice to select some form of a high grade of symmetry and consider what forms in systems of lower symmetry it limits. For instance, the isometric rhombic dodecahedron (Fig. 284), whose symbol is $\infty a : a : a$, is bounded by twelve planes, which correspond to two tetragonal, three orthorhombic, four monoclinic, and six triclinic forms, as is shown in the diagram at the top of the next page.

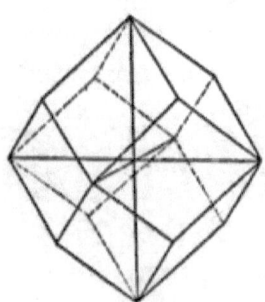

Fig. 284.

The limiting forms in different systems become of practical value in explaining the so-called "optical anomalies" exhibited by many crystals whose physical behavior is not strictly in accord with what seems to

178 CRYSTALLOGRAPHY.

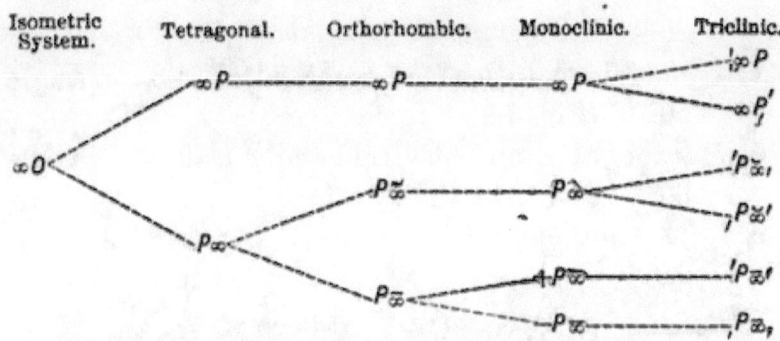

be their external form; and in this connection they will be again referred to in Chapter IX.

CLASSIFICATION OF ALL CRYSTAL FORMS BY THEIR SYMMETRY.

The following list of all the holohedral, hemihedral, tetartohedral and hemimorphic divisions of each system, classified according to their planes of symmetry, may prove of use. Roman numerals indicate principal planes of symmetry, and Arabic numerals secondary planes.

Nine planes of symmetry.
1. Isometric, holohedral. III + 6. (Fig. 51.)

Seven planes of symmetry.
2. Hexagonal, holohedral. I + 3 + 3. (Fig. 164.)

Six planes of symmetry.
3. Isometric, tetrahedral hemihedral. 6. (Fig. 94.)
4. Hexagonal, hemimorphic. 3 + 3. (Angle 30°.) (Fig. 237.)

Five planes of symmetry.
5. Tetragonal, holohedral. I + 2 + 2. (Fig. 126.)

Four planes of symmetry.
6. Tetragonal, hemimorphic. 2 + 2. (Angle 45°.) (Fig. 158.)

THE TRICLINIC SYSTEM.

Three planes of symmetry.

7. Isometric, parallel-face hemihedral. 3. (Angle 90°.) (Fig. 76.)
8. Hexagonal, rhombohedral hemihedral. 3. (Angle 60°.) (Fig. 194.)
9. Orthorhombic, holohedral. 3. (Angle 90°.) (Fig. 240.)

Two planes of symmetry.

10. Tetragonal, sphenoidal, hemihedral. 2. (Angle 90°.) (Fig. 155.)
11. Orthorhombic, hemimorphic. 2. (Angle 90°.) (Fig. 259.)

One plane of symmetry.

12. Tetragonal, pyramidal hemihedral. I. (Fig. 150.)
13. Hexagonal, pyramidal hemihedral. I. (Fig. 185.)
14. Monoclinic, holohedral. 1. (Fig. 265.)
15. Monoclinic, hemihedral. 1. (Fig. 275.)

No plane of symmetry.

16. Isometric, gyroidal hemihedral. (Fig. 66.)
17. Isometric, tetartohedral. (Fig. 109.)
18. Tetragonal, trapezohedral hemihedral. (Fig. 148.)
19. Tetragonal, trapezohedral tetartohedral.
20. Tetragonal, sphenoidal tetartohedral.
21. Hexagonal, trapezohedral hemihedral. (Fig. 183.)
22. Hexagonal, trapezohedral tetartohedral. (Fig. 221.)
23. Hexagonal, rhombohedral tetartohedral. (Fig. 234.)
24. Orthorhombic, sphenoidal hemihedral. (Fig. 256.)
25. Monoclinic, hemimorphic. (Fig. 276.)
26. Triclinic, holohedral. (Fig. 279.)

CHAPTER IX.

CRYSTAL AGGREGATES.

Kinds of Aggregates. In the preceding chapters we have considered only the crystal individual (p. 16) as a unit, and entirely independent of its relations to other individuals, either of the same or of other kinds. Crystals, however, often conform in their groupings to certain definite laws which therefore become an integral part of crystallography.

The ideal crystal individual is invariably a polyhedron whose interfacial angles are all less than 180°; the presence of re-entering angles on a crystalline surface therefore indicates the union of two or more individual crystals.

In a crystal aggregate (p. 16) the individuals may be all of one kind, or of different kinds. In the first instance the molecular arrangement may be completely parallel throughout all the individuals, which are then only separated from each other by their external planes. Such aggregates are called *parallel growths*. In other cases there may be no relation whatever between the molecular orientation of two contiguous individuals. Groups of crystals formed in this way are called *irregular aggregates*.

If we conceive of the resultant of all the attractive and repellent forces belonging to the crystal molecule as resolved into three components not at right angles, we can understand how the parallelism between the molecular structures of two individual crystals of the same substance may be partial. This has been already explained in Chapter I (p. 8, Figs. 5, 6, and 7). If all three axes of two similar molecules are parallel, the orientation is identical, as in parallel growths; if none of the axes are parallel, the orientation is wholly different, as in irregular aggregates; but if one or two of the three axes of one molecule are parallel to the corresponding axes of the other molecule, the orientation is neither complete nor wanting, but partial. Such a position as that last described may be imagined as produced by a revolution of one of a pair of completely parallel molecules, through an angle of 180° about either of its axes. Two crystal individuals which have this kind of a relative position are said to be *twins*.

In case the crystal aggregate is formed of individuals of different substances, there may be almost complete parallelism of orientation if the different crystals possess very nearly the same molecular structure. This is the case with many substances of analogous, though not of identical composition, which are called *isomorphous*. Such aggregates are therefore called *isomorphous growths*. In still other cases the crystals of one substance may, to a certain extent, affect the orientation of other crystals with a different molecular structure, which are deposited upon them. Such aggregates are called *regular growths*, to distinguish them from those in which the arrangement of the crystal individuals is entirely irregular.

The six possible categories of crystal aggregates may therefore be classified as follows:

182   CRYSTALLOGRAPHY.

I. HOMOGENEOUS AGGREGATES. Individuals of the same substance.
   1. Parallelism complete. . . *Parallel Growths.*
   2. Parallelism partial. . . . . *Twin Crystals.*
   3. Parallelism wanting. . *Irregular Aggregates.*

II. HETEROGENEOUS AGGREGATES. Individuals of different substances.
   4. Parallelism nearly complete.
                          Isomorphous Growths.
   5. Parallelism partial. . . . *Regular Growths.*
   6. Parallelism wanting.
                Irregular Heterogeneous Aggregates.

We shall now consider the essential characters of these six groups in succession.

I. AGGREGATES OF CRYSTALS OF THE SAME SUBSTANCE.

Parallel Growths. The molecular arrangement must be the same along all parallel lines within a crystal

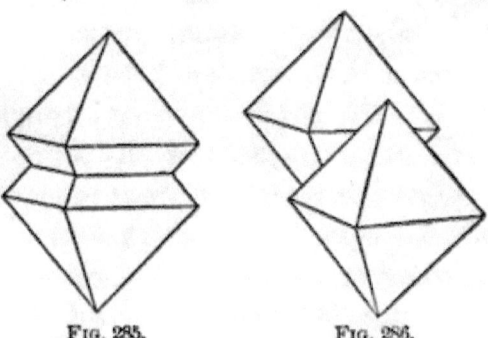

FIG. 285.   FIG. 286.

individual, but two or more individuals of the same substance may grow side by side, in such a position that their molecular arrangements are completely parallel throughout. *Such crystals must be sym-*

metrical with reference to some plane which is a plane of symmetry for each crystal form. They may, however, be united in this or in any other plane. This is shown in Figs. 285 and 286, which represent two octahedrons in exactly parallel position, and therefore symmetrical with reference to the faces of the cube.

In the first case, the cubic face is also the one in which the two individuals are united; while in the second, they are joined in an octahedral face, though their positions are still exactly parallel.

A large number of individuals may be united in this way where each is represented by only an extremely thin lamella (Fig. 287). This results in the alternate repetition of two planes, meeting at angles which are supplements of each other.

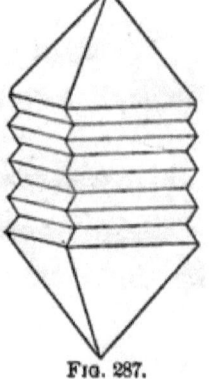

Fig. 287.

Strictly speaking, we should regard each re-entering angle as indicative of a separate individual, joined to the others by parallel growth. If, however, the width of the alternating planes is extremely small, the separation of the different individuals becomes very slight, and the effect is of a single crystal whose faces are finely striated. In such cases it is customary to speak of the alternating planes as producing an *oscillatory combination* of forms on a single individual. Striations caused in this manner are among the most frequent sources of imperfection on crystal planes, as will be more fully explained in the succeeding chapter.

Parallel growths become of importance in connection with the manner in which many substances crystallize. When the rate of increase is rapid, it not infrequently happens that a number of minute crystals group themselves in parallel position to form the skeleton of a larger crystal. In such cases the smaller forms have been called by Sadebeck *sub-individuals*. A familiar example of this mode of growth is offered by common salt; and the same thing may be observed in the case of sulphur, copper, gold, fluorspar, quartz and a variety of other substances.

Twin Crystals. (German, *Zwillinge*; French, *macles*, *hemitropes*.) When two crystals of the same substance, or two halves of the same crystal, are not in completely parallel position, they may still be joined in such a manner that some crystallographic plane, or at least some crystallographic direction, is common to both. In such cases the two crystals or halves of the same crystal are symmetrical with reference to some plane which *is not a plane of symmetry for the single individuals*. This is the most apparent distinction between parallel growths and what are known as crystals in twinning position, or *twin crystals*.

Symmetrical aggregates of this kind are, on account of their complexity, variety and frequency, of much importance in crystallography; and are therefore deserving of particular description.

The Twinning Plane and Twinning Axis. The relative position of two crystals in twinning position may be most readily understood by supposing that one has been revolved through 180° about some crystallographic direction, which thus remains common to both individuals. This direction or line of revolution is called the *twinning axis*, and it is in most cases normal to the plane with reference to which the two individuals become symmetrical after the revolution. This plane is called the *twinning plane*.

Suppose, for instance, that we imagine an octahedron cut into two equal parts parallel to an octahedral plane, and one half revolved 180° about a line normal to this plane. The result will be the position shown in Fig. 288, where the plane in which the octahedron is divided is the twinning plane, and the line normal to it is the twinning axis.

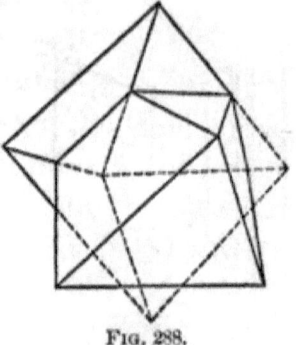

FIG. 288.
(Spinel.)

If, however, the octahedron were composed of a positive and negative tetrahedron in equal development, the two halves, after such a revolution, would not be symmetrical with reference to this plane, because a negative face would lie over a positive face, and *vice versa*. The two halves would then be symmetrical with reference to a face of the icositetrahedron, $2O2$, $\{211\}$. Such instances, where the twinning plane is not normal to the twinning axis, are not common.

Any crystallographic plane not a plane of symmetry may be a twinning plane, but the most frequent twinning planes in all systems are those which possess the simplest indices.

Contact Twins and Composition Face. Two individual crystals in twinning position are usually, though by no means always, united in a plane, which may or may not coincide with the twinning plane. Twins of this sort are called *contact* or *juxtaposition twins*, and the plane in which they are in contact is called their *composition face*. This, like the twinning plane, is generally a plane that has very simple indices.

Figs. 289 and 290 represent twins of two monoclinic

minerals, gypsum and orthoclase, in both of which the twinning plane is the orthopinacoid. In the first this is also the composition face, while in the second the composition face is at right angles to its twinning plane, and is itself the plane of symmetry, which could not therefore be a twinning plane.

Both twinning plane and composition face are crystallographically possible planes, except in the monoclinic and triclinic systems. Here they may be surfaces which are impossible as crystal planes, but they are either (1) perpendicular to a possible crystal edge, or (2) perpendicular to a possible face and parallel to a possible edge.

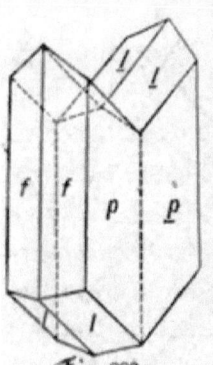

Fig. 289.
(Gypsum.)

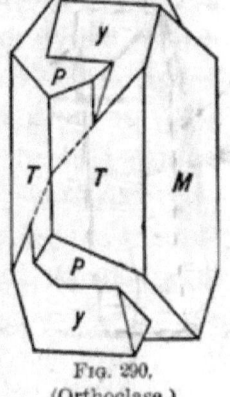

Fig. 290.
(Orthoclase.)

Penetration Twins. It happens quite frequently that two crystals in twinning position are not joined in a single plane, but that there is a complete interpenetration of both individuals. This is shown in Fig. 291, which represents two twin rhombohedrons, symmetrical with reference to a prism of the first order, but without any composition face. There is a complete interpenetration of their substance, and the space common to both is very irregularly distributed between them, as may be shown by an examination in polarized light. Only those portions of each crystal which project beyond the limits of the other possess a uniform

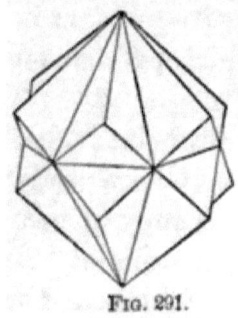

Fig. 291.
(Hematite.)

molecular structure. Twin crystals of this kind are called *penetration twins*.

Supplementary Twins. Partial forms (hemihedral, tetartohedral or hemimorphic) have lost more or less of the symmetry belonging to their corresponding holohedrons, and may consequently possess twinning planes which are impossible for the latter. These partial forms frequently form twins with the axes parallel for both individuals, which would, of course, be impossible for holohedral crystals. They are called *supplementary twins* (German, *Ergänzungszwillinge*), since the union, especially by penetration, of two hemihedrons in this manner, tends to restore the lost

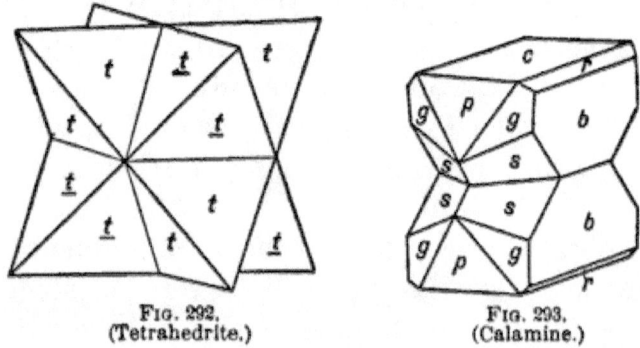

Fig. 292.
(Tetrahedrite.)

Fig. 293.
(Calamine.)

holohedral symmetry. Fig. 292 shows two tetrahedrons with parallel axes, and symmetrically placed with reference to the faces of the cube (tetrahedrite and diamond). Fig. 293 is another case of a supplementary twin, formed from two hemimorphic crystals, placed symmetrically to their basal pinacoid (calamine).

Repeated Twinning. To the second individual of a twin, a third may be placed in twinning position according to the same law as the first. This produces a *trilling* (German, *Drilling*). Four individuals related in this way make a *fourling* (German, *Vierling*), etc.

law may take place by either of two methods: (1) the twinning plane may remain parallel to itself, so that the alternate individuals are in parallel position; or (2) the twinning plane may change its direction, as when the two symmetrical faces of a prism successively serve as twinning planes. Figs. 294 and 295 illustrate these two methods of repeated twinning, where, in both cases, the unit prism is the twinning plane. The results of the first method are called *polysynthetic twins;* and of the second, *cyclic twins* (German, *Wendezwillinge*), on account of their tendency to produce circular groups, as in the case of rutile (Fig. 296). Such cyclic groups are more or less symmetrical, according as the angle between the successive sets of axes is more or less exactly a divisor of 360°.

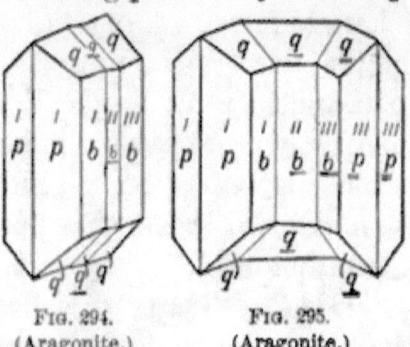

Fig. 294. (Aragonite.) Fig. 295. (Aragonite.)

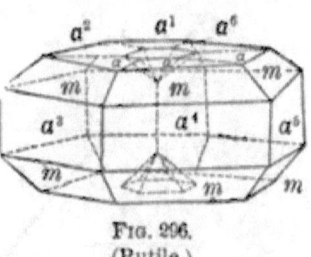

Fig. 296. (Rutile.)

Compound Twins. These are produced by the presence of two or more twinning laws in the same group. An instance of this is shown in Fig. 297, where two orthoclase twins, like that represented in Fig. 290, are again twinned parallel to their basal pinacoid. Compound twinning may lead to very complicated relations. Comparatively simple examples of it are found on crystals of the minerals marcasite, chalcocite, albite, staurolite, and tridymite.

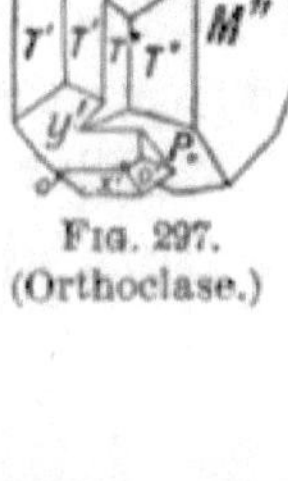

Fig. 297. (Orthoclase.)

Mimicry. The general tendency of twinning is to increase the grade of symmetry. Three orthorhombic individuals, the angle between whose symmetrical twinning planes is nearly 120°, may interpenetrate so as to form an almost perfectly hexagonal combination. This is the case with witherite ($BaCO_3$), the angle between whose twinning planes, $\infty P \wedge \infty P$, $\{110\} \wedge \{1\bar{1}0\}$, is 118° 30'; and with chrysoberyl, the angle between whose twinning planes, $P\breve{\infty} \wedge P\breve{\infty}$, $\{011\} \wedge \{0\bar{1}1\}$, is 119° 46' (Fig. 298). This group can also be explained by assuming the brachydome, $3P\breve{\infty}$, $\{031\}$ as twinning plane, whose interfacial angle $3P\breve{\infty} \wedge 3P\breve{\infty}$, $\{031\} \wedge \{0\bar{3}1\}$, is 59° 46'.

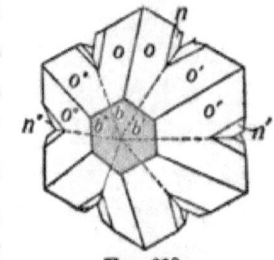

Fig. 298.
(Chrysoberyl.)

Pseudosymmetrical crystals (p. 92), that is, such as closely simulate a higher symmetry than they really possess, are enabled by repeated twinning to greatly increase their deceptive appearance. In many instances, their true character can only be determined by optical means. This phenomenon has been called by Tschermak *mimicry* (German, *Mimesie*). Examples of it are offered by the pseudo-monoclinic microcline and the pseudo-rhombohedral chabazite, which are really triclinic; by the pseudo-tetragonal apophyllite, which is monoclinic; and by the pseudo-isometric leucite, which, at ordinary temperatures, is orthorhombic. Harmotome and phillipsite are monoclinic, but they possess interfacial angles which allow them, by repeated twinning, to closely approach a tetragonal, or even an isometric habit. This property of mimicry is of importance in explaining many of the so-called "optical anomalies" exhibited by crystals.

Mode of Formation of Twin Crystals. As has been already stated in Chapter I, the perfection of crystallization in a given substance is inversely proportional to the rapidity of its solidification. If the molecular forces have full time to act, the parallelism of the molecules will be complete. In case the hypothesis that twin crystals are due to a partial parallelism of the molecules is correct, we might expect to produce them by artificially retarding the motion of the molecules at the time of solidification. This has actually been accomplished by O. Lehmann, who found that barium chloride and some other salts, which habitually produce simple forms on crystallizing from an aqueous solution, appeared in twins when such a solution was mixed with gum or some other viscous substance which retarded molecular movement.*

In the case of contact twins, the hemitropic character appears to date from the first inception of the crystal, as is shown by the fact that the most minute individuals observable with the microscope are as perfect twins as those of large size. When such a double crystal has once been started, its growth is regular in both directions away from the composition face. The result of this is a general shortening of each individual in the line of the twinning axis, and the frequent production of two half-crystals in twinning position.

Penetration or repeated twins, on the other hand, show a constant tendency to the addition of layers in hemitropic position with reference to those which precede. This may be due to the impurity of solution or viscosity of the magma in which crystallization is taking place.

---

* Molecularphysik, vol. I. p. 415.

Mimetic twins may be produced in dimorphous substances by physical conditions unsuited for the existence of the molecular arrangement with which they solidified. The new conditions may produce a new molecular structure, which, by twinning, is made to fit into the original crystal form. This is the case with boracite, tridymite and leucite. When the original conditions are restored (generally by raising the temperature) the internal structure again comes into accord with the external form.

Examples of Common Twinning Laws.* It will be found advantageous to briefly explain a number of figures which represent concrete examples of the commonest modes of twinning in each of the six crystal systems.

In the isometric system, a large majority of all twins are formed according to the law illustrated by Fig. 288 (p. 185), where the twinning plane is an octahedral face, and the twinning axis a normal to this. This law, commonly called the "spinel law," from its frequent occurrence on crystals of the aluminate, spinel, is capable of producing a great variety of results which differ according to the habit of the crystals which are twinned, as well as according to whether contact or penetration twins are formed. Figs. 299, 300 and 301 show contact twins of the three simplest isometric holohedrons according to this law; while Figs. 302, 303 and 304 represent penetration twins of the same forms by the same method. Fig. 305 shows

---

* For a more complete illustration of this subject see Rose-Sadebeck's Crystallography, vol. II (1876); Klein's Dissertation on Crystal Twinning and Distortion (1876); and E. S. Dana's Text-book of Mineralogy.

a penetration twin of two tetrahedrons, also symmet-

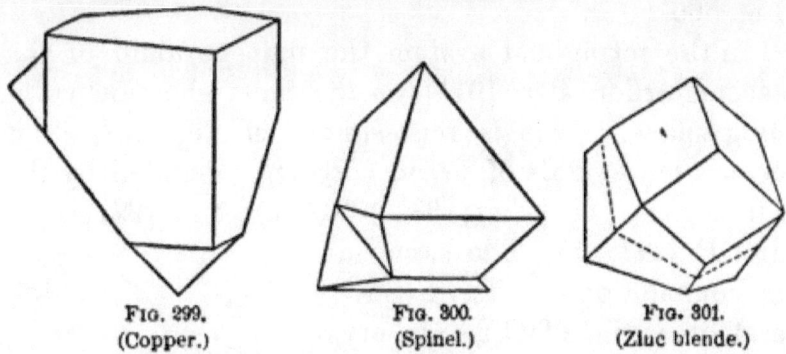

Fig. 299. (Copper.)   Fig. 300. (Spinel.)   Fig. 301. (Zinc blende.)

rical to the octahedral face; while another penetration twin of the same two forms symmetrical to the

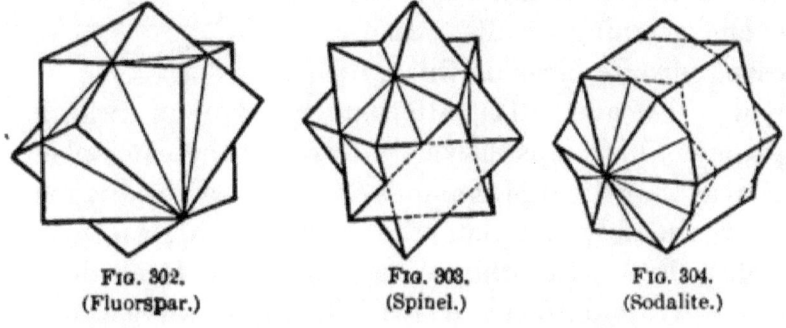

Fig. 302. (Fluorspar.)   Fig. 303. (Spinel.)   Fig. 304. (Sodalite.)

face of the cube (which is no longer a plane of symmetry in these figures) is shown in Fig. 292 on p. 187.

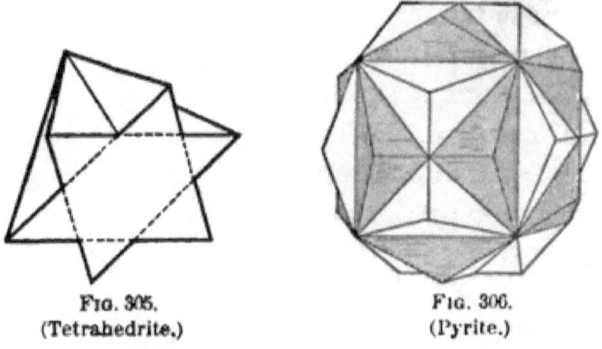

Fig. 305. (Tetrahedrite.)   Fig. 306. (Pyrite.)

A penetration twin of two pentagonal dodecahedrons (known as the iron cross) whose twinning plane is the rhombic dodecahedron, ∞O, $\{110\}$, is shown in Fig. 306.

In the tetragonal system the unit pyramid of the second order, $P\infty$, $\{011\}$, is the most common twinning plane. This is represented in Fig. 307, as it occurs on crystals of zircon ($ZrSiO_4$), bounded by the forms ∞P, $\{110\}$ (m); $3P$, $\{331\}$ (u); $2P$, $\{221\}$ (p); and P, $\{111\}$ (o). The same law is common on cassiterite (SnO_2), and on rutile (TiO_2). On crystals of the latter mineral repeated twinning is common. This is accomplished in two ways where cyclic twinning results. In one case, *opposite* faces of the pyramid, (011) and (0$\bar{1}$1), alternately serve as twinning planes. This keeps the vertical axes in the same plane, and, as they diverge at angles of 65° 35′, six individuals approximately complete the circuit, as shown in Fig. 296 (p. 188). In other cases, *contiguous* faces of the same pyramid (011) and (101) alternately serve as twinning planes. This causes the vertical axes to form a zigzag, and eight individuals are necessary to com-

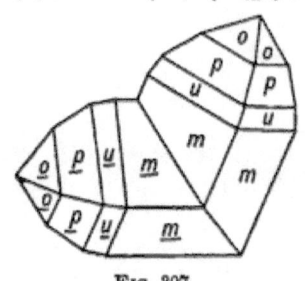
Fig. 307.
(Zircon.)

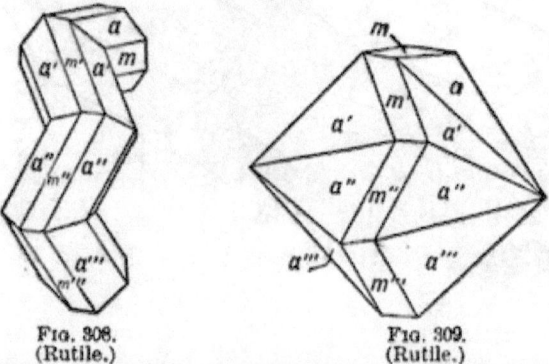
Fig. 308.
(Rutile.)
Fig. 309.
(Rutile.)

plete 360° (Fig. 308). Rutile twins of this kind from Graves Mt., Ga., often have their prism faces of the second order, $\infty P \infty$, {010} (*a*), extended to intersection, which produces a tetragonal scalenohedron, as shown in Fig. 309.

In the tetragonal mineral chalcopyrite (FeCuS$_2$),

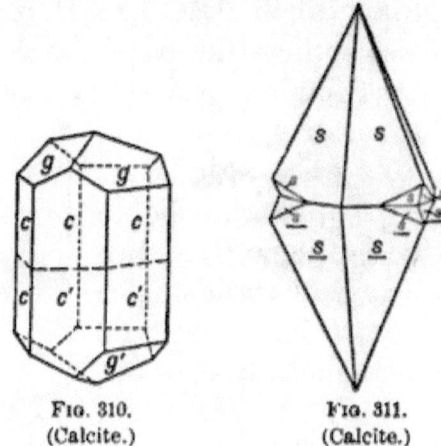

Fig. 310.
(Calcite.)

Fig. 311.
(Calcite.)

which approaches very closely to an isometric crystallization (p. 92), the twinning plane is the unit pyramid, P, {111}. This corresponds to the spinel law in the isometric system (p. 191). The tetragonal but pyram-

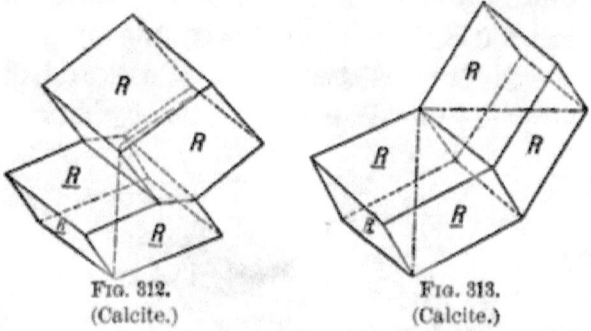

Fig. 312.
(Calcite.)

Fig. 313.
(Calcite.)

idally hemihedral scheelite (CaWO$_4$) forms supplementary twins, where the twinning plane is the unit prism, ∞P, {110}.

In the hexagonal system holohedral substances and consequently holohedral twins are rare. On pyrrhotite, magnetic pyrite (Fe$_7$S$_8$), the unit pyramid,

P, {111}, has been observed as twinning plane; and on tridymite (SiO$_2$), the two pyramids, $\frac{1}{6}P$, {10$\bar{1}$6} and $\frac{3}{4}P$, {30$\bar{3}$4}. On rhombohedral crystals twinning is very common, particularly so on the best representative of this group, calcite (CaCO$_3$). Here we have as twinning plane sometimes the basal pinacoid with the vertical axis as twinning axis (Figs. 310 and 311); sometimes the negative rhombohedral face, $-\frac{1}{2}R$, $\kappa\{01\bar{1}2\}$ (Fig. 312); sometimes the positive rhombohedron, R, $\kappa\{10\bar{1}1\}$, bringing the two vertical axes nearly at right angles to one another (Fig. 313); and sometimes the rhombohedron, $-2R$, $\kappa\{02\bar{2}1\}$ (Fig. 314). A penetration twin of two rhombohedrons, as sometimes seen on crystals of ferric oxide, hematite, is shown in Fig. 291 (p. 186).

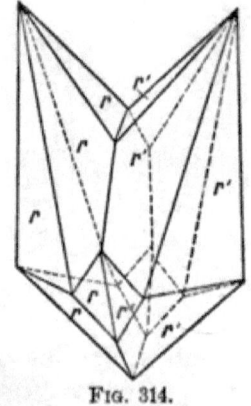

FIG. 314.
(Calcite.)

Tetartohedral hexagonal forms, like those occurring on crystals of quartz, produce supplementary twins, which tend to restore a higher grade of symmetry. Thus a complete interpenetration of two right-handed or of two left-handed individuals, one of which has been revolved 180° (or 60°) about its vertical axis, reproduces a trapezohedral hemihedral form (Dauphiné law, Fig. 315); while a similar interpenetration of a right- and a left-handed crystal restores a scalenohedral symmetry, with the prism of the second order, $\infty P2$, $\kappa\tau\{11\bar{2}0\}$ as twinning plane (Brazilian law, Fig. 316).

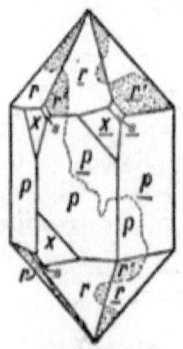

FIG. 315.
(Quartz.)

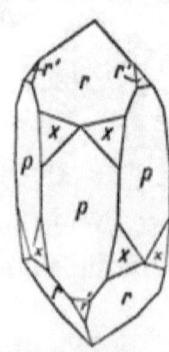

FIG. 316.
(Quartz.)

Contact-twins of quartz, with the individuals symmetrical to both ∞P, $\{10\bar{1}0\}$ and R, $\kappa\tau\{10\bar{1}1\}$ also occur.

In the orthorhombic system the most common twinning plane is the unit prism ∞P, $\{110\}$, as exemplified by both the polysynthetic and cyclic twins of aragonite shown in Figs. 294 and 295 (p. 188). Tabular crystals of lead carbonate (cerussite) bounded by the forms $\infty P\breve{\infty}$, $\{010\}$ (b); ∞P, $\{110\}$ (m); and P, $\{111\}$ (p) (Fig. 317), form groups in which both ∞P, $\{110\}$ and $\infty P\breve{3}$, $\{130\}$ act as twinning planes. Orthorhombic iron disulphide (marcasite) sometimes shows cyclic groups of five individuals, bounded by ∞P, $\{110\}$ (M); $P\breve{\infty}$, $\{011\}$ (l), and $0P$, $\{001\}$, and united into a pentagonal figure by ∞P, $\{110\}$ (M) as twinning plane (Fig. 318). The closely related iron sulpharsenide (arsenopyrite), bounded by ∞P, $\{110\}$ (M),

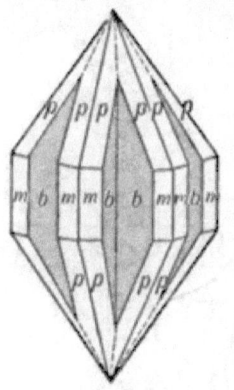

Fig. 317.
(Cerussite.)

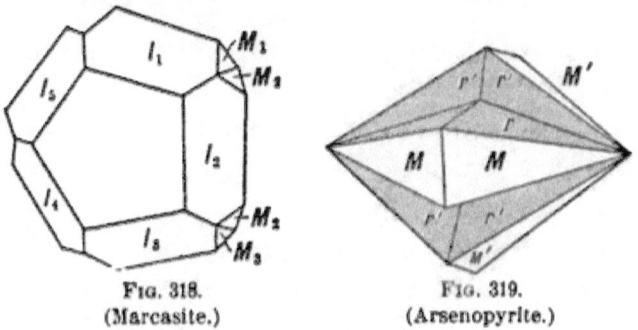

Fig. 318.
(Marcasite.)

Fig. 319.
(Arsenopyrite.)

and $\tfrac{1}{3}P\breve{\infty}$, $\{013\}$ (r), forms penetration twins of two individuals, where $P\breve{\infty}$, $\{101\}$ acts as twinning plane (Fig. 319).

The group of chrysoberyl crystals shown in Fig. 298 (p. 189) are united with $3P\breve{\infty}$, $\{031\}$ as twinning plane. The copper sulphide, chalcocite, forms three kinds of twins symmetrical to ∞P, {110}, to $\tfrac{1}{3}P\check{\infty}$, {043}, and to $\tfrac{1}{2}P$, {112} respectively. The latter is shown in Fig. 320. The orthorhombic silicate, staurolite, which is commonly bounded by the planes $\infty P\check{\infty}$, {010} (*o*); OP, {001} (*P*); ∞P, {110} (*M*); and $P\check{\infty}$ {101} (*r*), forms rectangular crosses with $\tfrac{3}{2}P\check{\infty}$, {032} as twinning plane (Fig. 321), and also

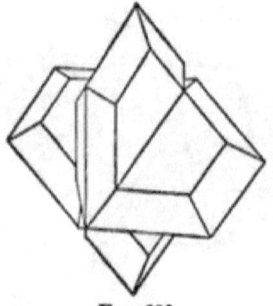

FIG. 320.
(Chalcocite.)

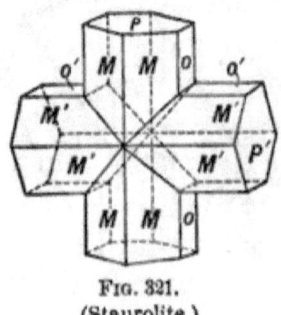

FIG. 321.
(Staurolite.)

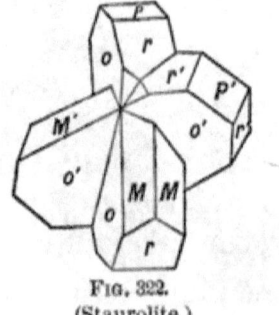

FIG. 322.
(Staurolite.)

oblique crosses with $\tfrac{3}{2}P\tfrac{\bar{3}}{2}$ {232} acting in the same capacity (Fig. 322).

In the monoclinic system any face may serve as twinning plane except the clinopinacoid, which is the plane of symmetry. Even this, however, is not infrequently a composition face for contact twins, as in the case of orthoclase (Fig. 290, p. 186). The most common twinning plane is the orthopinacoid $\infty P\check{\infty}$, {100}, as may be seen in the case of gypsum (Fig. 289, p. 186) and malachite (Fig. 323). The same law is also exemplified in augite, hornblende, feld-

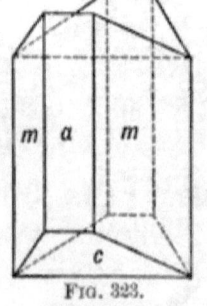

FIG. 323.
(Malachite.)

spar, and epidote. A twin of the latter mineral with the forms OP, $\{001\}$ (M); $\infty P\bar{\infty}$, $\{100\}$ (T); $P\bar{\infty}$, $\{101\}$ (r); $\infty \check{P}\infty$ $\{010\}$ (P); $+P$, $\{\bar{1}11\}$ (n); $\infty \check{P}2$, $\{210\}$ (u); and $\frac{1}{2}\check{P}\infty$, $\{012\}$ (k), is shown in Fig. 324. The monoclinic feldspar, orthoclase, exhibits a number of different twinning laws. In addition to the one explained on p. 186 (Carlsbad law), the twinning plane is sometimes the basal pinacoid

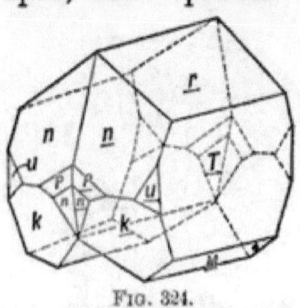

Fig. 324.
(Epidote.)

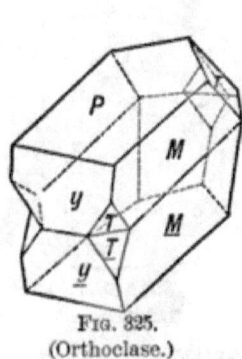

Fig. 325.
(Orthoclase.)

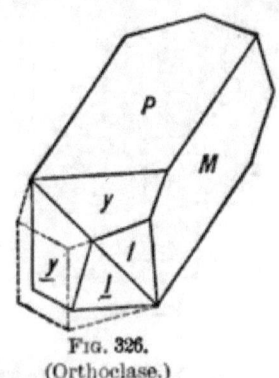

Fig. 326.
(Orthoclase.)

(Mannebacher law); and sometimes a clinodome whose symbol is $2\check{P}\infty$, $\{021\}$ (Baveno law). Contact twins formed according to these two laws are shown in Figs. 325 and 326. The silicate, augite, sometimes has as twinning plane the clinopyramid $\check{P}2$, $\{122\}$ (Fig. 327). The monoclinic mica differs but slightly from a hexagonal mineral in its symmetry and angles. Its twinning axis is often a line in the basal pinacoid, normal to the combination edge OP, $\{001\}$: ∞P, $\{110\}$. The

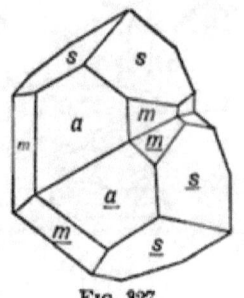

Fig. 327.
(Augite.)

CRYSTAL AGGREGATES.

twinning plane is therefore a crystallographically impossible face, normal to the basal pinacoid; while the composition face is generally the basal plane (Fig. 328).

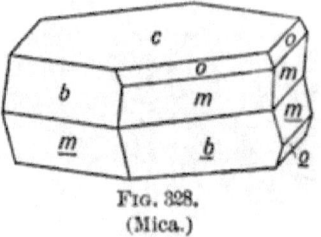

Fig. 328.
(Mica.)

In the triclinic system any face may act as twinning plane, or may be, and frequently is one which, on account of its irrational indices, is not possible as a crystal plane. On crystals of the soda

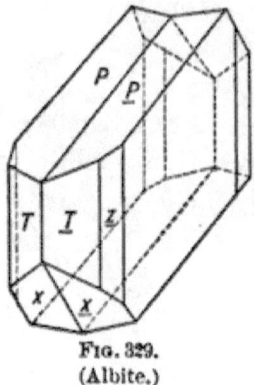

Fig. 329.
(Albite.)

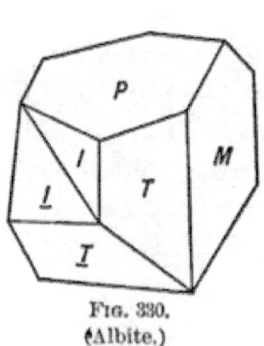

Fig. 330.
(Albite.)

feldspar, albite, we find the brachypinacoid as twinning plane (Fig. 329), and also the brachydome,

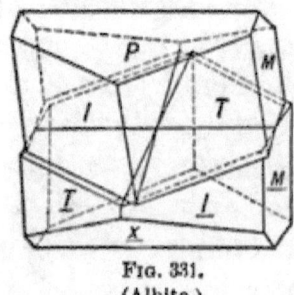

Fig. 331.
(Albite.)

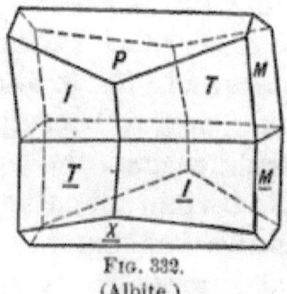

Fig. 332.
(Albite.)

$,2P\bar{\infty}'$ {021} (Fig. 330). The analogue of the Mannebacher law occurs on the triclinic feldspars when the twinning axis is the macrodiagonal, and the twinning plane normal to this, and therefore no possible crystal plane (Pericline law, Fig. 331). To avoid the re-entrant angles produced by this law when the basal pinacoid is the composition face, the two individuals are generally united in that particular *rhombic section* (German, *rhombischer Schnitt*) which is common to both (Fig. 332). The exact position of this section, of course, depends on the relative inclinations of the axes. A variety of twinning laws, where the twinning planes are not crystallographically possible faces, have also been observed on the triclinic orthosilicate of alumina, Al_2SiO_5 (cyanite, disthene).

Irregular Homogeneous Aggregates of Crystals. When crystallization begins simultaneously at many separated points in a saturated solution or cooling magma, individual crystals are formed whose molecular structures are entirely independent in their orientation, although they may be identical in their nature. If such crystals continue to grow until they come in contact, a wholly irregular aggregate is the result (Figs. 21 and 22, p. 17). Such aggregates are classified according to the extent to which their crystals are individually developed (crystal and crystalline aggregates); according to their texture (porous or compact); and according to the size of their grain (coarse or fine).

Other crystalline aggregates are not entirely irregular in the arrangement of their component individuals. Certain acicular crystals have a tendency to arrange themselves radially about a point, so as to form spherulitic groups (wavellite, stilbite), while others have a similar radial arrangement about a line (aragonite, "flos ferri"). Many fibrous crystals group themselves with their long axes nearly parallel (gypsum, chrysotile, asbestos). Tabular or scaly crystals form lamellar or foliated aggregates (wollastonite, brucite, gypsum, mica).

These types of aggregates produce an almost endless number of varieties and compound forms, whose special description must be sought in a larger work.

In still other cases there is a near approach to parallelism in the individual crystal forming the group. Examples of this may be found in the so-called "iron-rose" (hematite) of Switzerland; in stilbite, and in the twisted quartz crystals from Switzerland.

II. Aggregates of Crystals of Different Substances.

Isomorphous Growths. It was long ago observed by Mitscherlich that substances of analogous chemical composition were apt to possess very similar crystal forms. Such substances he called *isomorphous*. The shape, size and mode of arrangement of their physical molecules must be nearly alike, since the test of isomorphism is the ability of molecules of two or more substances to enter indiscriminately into the formation of a single crystal; or at least the ability of a crystal of one substance to continue its growth in a saturated solution of another. If we suspend a crystal of chromium alum in a solution of potash alum, it will soon be coated with a transparent layer of the latter salt, which perfectly preserves the form of the original crystal. This may be again covered with a similar layer of chromium alum, and so on to any number of successive zones whose physical molecules are all in parallel orientation. The same thing takes place when a crystal of calcium carbonate is hung in a solution of sodium nitrate. Such concentric zones of different substances with completely parallel molecular structures are of frequent occurrence in nature, but they are only to be found among isomorphous compounds, like the garnets, tourmalines, micas, pyroxenes, feldspars, etc. A striking example is the parallel growth between crystals of the isomorphous xenotime (YPO$_4$) and malacon (ZrSiO$_4$ + aq). Another is the growth of a zone of epidote around the isomorphous silicate, allanite (Fig. 333).

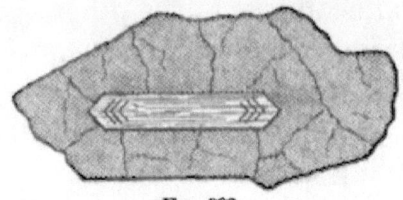

Fig. 333.

Regular Growths of Different Minerals. It is worthy of note that a certain regularity of arrangement exists between the crystals of substances which are altogether unlike in chemical composition. For instance, the triclinic cyanite and the orthorhombic staurolite grow together so that their crystals have one face and one axis in common. The tetragonal rutile (TiO$_2$) grows upon the rhombohedral titanic iron (FeTiO$_3$), so that its prism of the second order, $\infty P \infty$ (100), coincides with the basal pinacoid of the latter mineral, while its vertical axis has the direction of one of the intermediate lateral axes of the iron ore (Fig. 334). Similar examples of partial orientation are to be found between quartz and calcite; between chalcopyrite and tetrahedrite; and between magnoferrite and hematite.

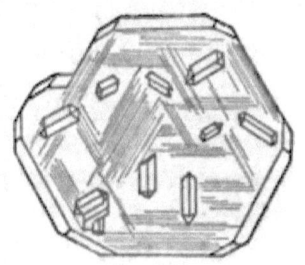

Fig. 334.

They also exist between dimorphous modifications of the same substance, as marcasite and pyrite, calcite and aragonite, pyroxene and hornblende, etc., where they have originated, in some cases at least, by the partial alteration of one modification into the other.

In other cases, the crystals so related have analogous compositions and crystallize in different systems, but with very similar forms, as orthoclase and albite, orthorhombic and monoclinic pyroxene, etc.

A partial parallelism in orientation is also frequent between the inclusions in many crystals and their host, as for instance, rutile in mica, augite in leucite, hematite in feldspar, coaly matter in andalusite (chiastolite), etc. This subject becomes of considerable importance in microscopical mineralogy and petrography.*

Irregular Heterogeneous Growths of Crystals. This is the most general form of grouping possible. It is purely accidental and obeys no rules, although a thorough understanding of it implies a knowledge of the laws of association and paragenesis, which are of great importance in the mineral world. The study of heterogeneous crystal aggregates belongs, however, rather to the domain of petrography than to that of mineralogy or crystallography.

---

* For full description and list of regular growths of minerals of different species, both isomorphous and otherwise, see Sadebeck: Angewandte Krystallographie, pp. 244–249; and O. Lehmann: Molecularphysik, vol. I. pp. 293–407. The latter work cites numerous instances from artificial salts.

CHAPTER X.

IMPERFECTIONS OF CRYSTALS.

Sources of Imperfection. We have thus far considered crystals in their ideal development; that is, as symmetrical polyhedrons, bounded by mathematically plane surfaces intersecting at fixed angles. If the molecular forces of a single substance at the time of its solidification were entirely free to act, without hindrance of any kind; and if the crystal growth could proceed with great slowness and perfect regularity in all directions, such ideal crystal forms would undoubtedly result. So sensitive, however, are these forces to obstacles of many kinds that it is only in rare cases, either in nature or the laboratory, that they succeed in accomplishing the most perfect result of which they are capable.

An acquaintance with the theoretical crystal form and surface may be obtained from figures and models constructed so as to show them in their ideal perfection. Natural crystals, however, rarely attain such perfection of development, and a knowledge of some of the causes which render them more or less incomplete is necessary for their thorough understanding.

Crystals may fall short of perfection by—
    1. Distortion of form.
    2. Irregularity of planes or angles.
    3. Internal impurity.

IMPERFECTIONS OF CRYSTALS.

Distortion of Crystal Form. The most common source of this is unequal rapidity of crystal growth in different directions. This may result in a disguising of the real form, as has been already described and illustrated in Chapter I. When carried to excess, certain planes of a form are altogether crowded out (merohedrism, p. 39), which often produces close imitations of shapes characteristic of other systems.* Instances of such distortion are particularly frequent in the isometric system, where elongation in the direction of a principal axis produces a tetragonal; in the direction of a trigonal axis (p. 47), a rhombohedral; and in the direction of a digonal axis, an orthorhombic appearance. This may be seen from the three following figures of the rhombic dodecahedron distorted in these three directions (Figs. 335, 336 and 337). Fig. 338 gives the result of an elongation of the icositetrahedron, 2O2, {211} in the direction of the trigonal axis.

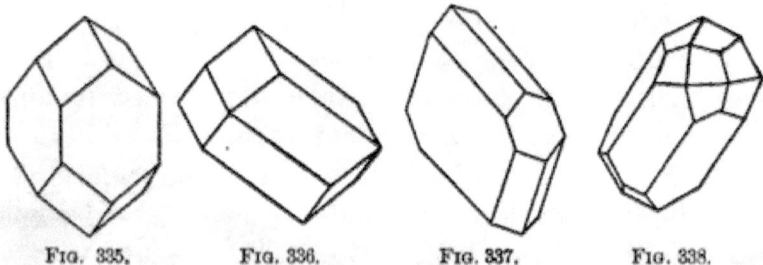

FIG. 335.   FIG. 336.   FIG. 337.   FIG. 338.

When accompanied by merohedrism, which generally obeys some definite law, surprising results are sometimes produced by the distortion of isometric forms.

---

* This phenomenon has been called *pseudosymmetry* by Sadebeck, who used the term in a sense very different from that in which it is employed by Tschermak (see p. 92). In the former usage the imitation is always of a lower, while in the latter it is of a higher grade of symmetry.

The octahedron, by the crowding out of two of its opposite faces, may become a rhombohedron (Fig. 339). Certain crystals of green fluorspar from Saxony, showing the form $\infty\, O3$, $\{310\}$, have one half of their planes developed at the expense of the other half, so as to produce a hexagonal scalenohedron (Fig. 340). The icositetrahedron is peculiarly liable to such distortion.

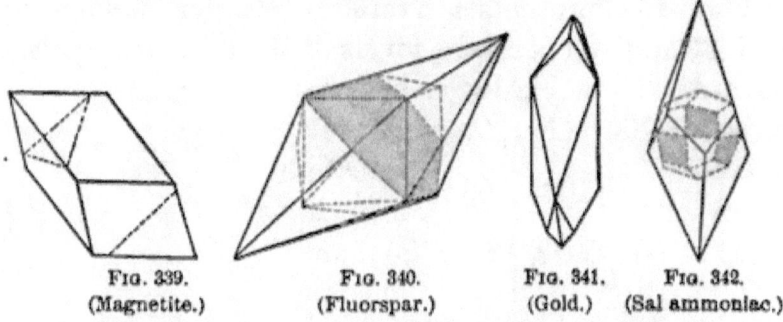

FIG. 339. (Magnetite.)   FIG. 340. (Fluorspar.)   FIG. 341. (Gold.)   FIG. 342. (Sal ammoniac.)

Gold crystals of the form $3O3$, $\{311\}$, sometimes resemble combinations of rhombohedron and scaleno-

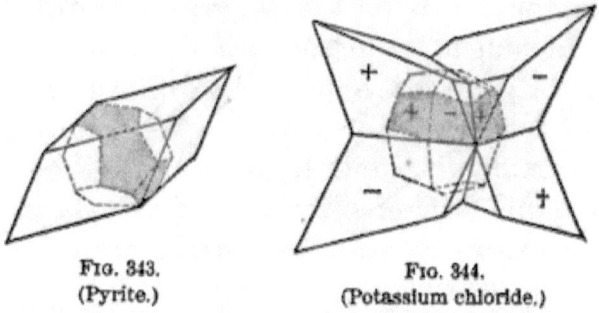

FIG. 343. (Pyrite.)   FIG. 344. (Potassium chloride.)

hedron (Fig. 341). The same form on ammonium chloride may have only six of its twenty-four planes developed, and in this way give rise to a form resembling a tetragonal trapezohedron (Fig. 342).

It has been observed that the pentagonal dodecahedron, $\left[\dfrac{\infty\, O2}{2}\right]$, $\pi\,\{201\}$, on pyrite may have but six of its twelve planes developed in such a manner as to produce an apparent rhombohedron (Fig. 343);* and the icositetrahedron, 4*O*4, {411}, on potassium chloride sometimes produces a similar result by the survival of only one fourth of its planes. In fact, two such rhombohedrons, in apparent twinning position, may be derived from the planes of the same icositetrahedron (Fig. 344).†

Fig. 345 represents a crystal of iron pyrites showing representatives of all the forms of the isometric system except the rhombic dodecahedron:
∞ *O*∞ {100} (*P*); *O*, {111} (*d*); $\left[\frac{\infty\, O2}{2}\right]$, π{102} (*e*); $\left[\frac{4O2}{2}\right]$, π{214} (*s*); 2*O*2, {211} (*o*); and 3*O*, {133} (*t*). It is shortened in the direction of one of the principal axes, which, together with the fact that only two of the three

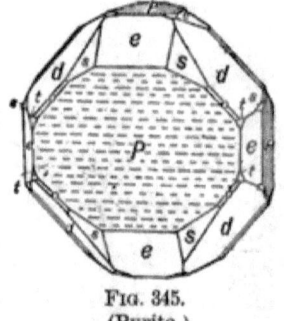

Fig. 345.
(Pyrite.)

faces belonging to the forms *s*, *o* and *t* are developed in each octant, gives to it a decidedly orthorhombic habit.

It has recently been shown to be not improbable that the mineral acanthite, long recognized as the orthorhombic form of silver sulphide, is only a distortion of the more common isometric form of the same substance argentite.

While a distortion of the perfectly symmetrical crystal form by elongation or flattening is often without any apparent cause, the same result is, in many other cases, produced by evident hindrances to growth in certain directions. For instance, such minerals as

---

* Neues Jahrbuch für Mineralogie, etc., 1889, II. p. 260.

† A. Knop: Molecularconstitution und Wachsthum der Krystalle (1867, p. 50).

garnet, tourmaline or quartz diverge from their usual habit and crystallize in the thinnest possible plates when they are formed in mica, whose perfect cleavage allows of their development most readily along one plane.

Other imperfections of crystal form are due to the action of external mechanical forces, which cause bending or breaking. Ductile substances, like the metals, or soft substances, like gypsum or stibnite, are peculiarly liable to the first of these distortions; although such brittle minerals as apatite or quartz, when imbedded in crystalline limestone, are frequently found to be bent without breaking. In many other cases the crystal is broken and its fragments more or less displaced by movements in its matrix. Crystals imbedded in granite veins, like tourmaline or beryl, frequently exhibit this phenomenon.

The most common imperfections in crystal form are due to the manner of their attachment to the surface on which they rest. One end of an individual is usually prevented in this way from assuming its characteristic planes. Crystals that are bounded on all sides by their own faces are of comparatively rare occurrence.

Imperfections of Crystal Planes. Theoretically even, perfectly reflecting crystal planes are almost as much the exception as the ideally developed crystal forms. Irregularities may be produced on crystal planes by (1) striation, (2) curvature, (3) uneven growth, (4) corrosion.

(1) *Striation.* A parallel striation of a crystal face may be produced by the union of many individuals in either parallel or reversed position. The first produces what was described in the last chapter (p. 183) as an

oscillatory combination of two contiguous planes, which are alternately developed. The horizontal striation, so common on the prismatic faces of quartz crystals, is due to the alternate development of the prism and a steep rhombohedron. Striation of this kind must always accord with the symmetry of the crystal, and often shows the hemihedral nature of a crystal upon which no real hemihedral face appears. Cubes of iron pyrites frequently show a striation of their planes

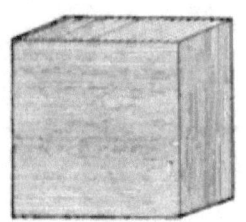

FIG. 346.
(Pyrite.)

in *one* direction, which is perpendicular to the striation on all contiguous faces, owing to oscillatory combination with the pentagonal dodecahedron (Fig. 346). This would not be possible in a holohedral cube.

Striation of crystal planes may also be produced by repeated polysynthetic twinning (p. 188). This is well illustrated in the case of the triclinic feldspar, albite (Fig. 347). The twinning plane is here the brachypinacoid, and the contact of a large number of fine lamellæ, alternately in twinning position, would evidently produce a striation on the basal plane parallel to the brachydiagonal axis. Striations due to this cause are common on crystals of pyroxene, calcite, sphene, aragonite, and many other minerals. The direction of the striation of course depends in each case on what face acts as twinning plane. In some instances two or more sets of parallel lamellæ are intercalated parallel to different crystal faces which may or may not belong to the same form;

FIG. 347.
(Albite.)

that is, they may be produced by the same, or by different twinning laws.

(2) *Curvature of Crystal Planes.* This is quite a constant property of some substances. It may be due to a very fine oscillatory combination or to an irregularity of growth and a want of perfect parallelism between sub-individuals. A common instance is the diamond, whose faces except those of the octahedron, are almost always curved (Fig. 348). Crystals of calcite and gypsum are often curved (Fig. 349), while rhombohedrons of magnesian calcium carbonate (dolomite) are sometimes distorted into saddle-like forms (Fig. 350). The

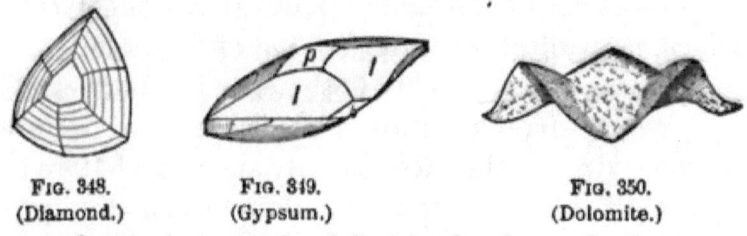

FIG. 348. (Diamond.)   FIG. 349. (Gypsum.)   FIG. 350. (Dolomite.)

twisted quartz crystals of Switzerland are also instances of this kind.

(3) *Irregularities of growth* often produce unevenness of crystal planes. Many faces break up, especially near their combination-edges, into other planes having

FIG. 351. (Quartz.)

very nearly, but not quite, their own position. These are called *vicinal planes* (p. 26). Sometimes these vicinal planes are developed in the form of very flat pyramidal protuberances with more or less curved edges and faces. These are very common on the rhombohedral faces of quartz (Fig. 351), but occur to a greater or less extent on the crystals of most other substances.

IMPERFECTIONS OF CRYSTALS. 211

In other cases, uneven planes are due to incomplete growth. Many crystals whose growth is rapid tend to form skeletons, by arranging sub-individuals along the axes and edges, and sometimes these skeletons do not become entirely filled up. A common instance of this is seen in the hopper-shaped faces of salt (NaCl) crystals, and in the artificial crystals of lead sulphide (Fig. 352).

Fig. 352. (Galena.)

A drusy appearance of crystal planes is produced by the projection of sub-individuals above the average surface, as is often seen in the case of fluorspar.

(4) *Corrosion* subsequent to the formation of crystals may produce irregularities of their surfaces. If this action is extreme, the faces usually succumb before the edges and angles, so that skeleton forms, closely resembling growth forms, may result. If the solvent action is less, natural etched figures are produced, which well display the crystals' true symmetry.

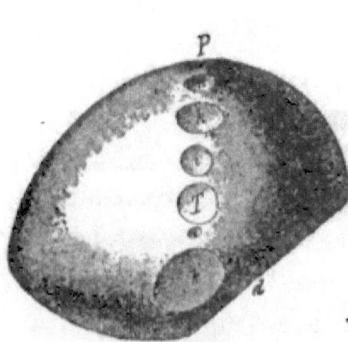

Fig. 353. (Olivine.)

A rounding-off of edges and angles and the production of a "glazed" appearance, as though there had been a partial fusion, is very common on all crystals which occur in crystalline limestone, as, for instance, quartz, calcite, pyrite, galena, apatite, chondrodite, feldspar, pyroxene, hornblende, tourmaline, scapolite, zircon, sphene, and spinel.

This is doubtless the result of some chemical action, although it has never been satisfactorily explained. In some cases, the quartz crystals of Herkimer Co., N. Y., have their edges so worn that only minute rounded areas of their crystal planes remain ; and the preceding figure (353) shows a similar development of an olivine crystal observed by Rose, where the planes of a single zone are developed as circular surfaces.

All imperfections of crystal surfaces express the symmetry of the form. On combinations, some faces are dull, while others are bright ; some are striated, while others are smooth ; some are uneven, while others are even ; but *all crystallographically equivalent planes are similarly affected.*

False Planes. Apparent crystal faces, whose position is not that of true crystal planes, may be produced by oscillatory combination, as in the case of tapering quartz crystals ; or by contact, during crystal growth, with some smooth surface, as in the case of the so-called "Babel quartz."

Variation in Crystal Angles. Most distortions of form do not at all affect the interfacial angles. But even these are in some cases observed to vary, though generally only within narrow limits. Such differences in angles exhibited by crystals of the same substance are to be accounted for (1) by slight differences in chemical composition ; (2) by variations of temperature ; (3) by mechanical action, either during or subsequent to the crystal's formation ; (4) by change of molecular arrangement through paramorphism or pseudomorphism. Some species appear to be much more sensitive to these agencies than others, and therefore more frequently exhibit variations in their interfacial angles.

Internal Impurities of Crystals. These may consist of (1) intermolecular substance in the form of a dilute pigment; (2) gas inclusions; (3) fluid inclusions; (4) glass inclusions; (5) inclusions of unindividualized matter; (6) crystals of other substances. Such inclusions may also be (1) regularly or (2) irregularly arranged; and (1) of primary or (2) of secondary origin.

(1) *Dilute pigment.* Many crystals are variously colored by minute quantities of matter scattered molecularly through them. The nature of this pigment is generally indeterminable. Remarkable examples are presented by crystals of corundum and tourmaline. The color may often be affected by temperature. The yellow Brazilian topaz may be made permanently pink by heating; while the green microcline (Amazon stone) and the smoky quartz (cairngorm) may be decolorized by the same means.

(2) *Gas inclusions.* These are common in crystals of both aqueous and igneous origin. Rose found in crystals of rock-salt inclusions of marsh-gas and hydrogen. The gas imprisoned in the rock-salt of Wieliczka expands violently on heating (*Knistersalz*). Carbon dioxide occurs in the quartz of many granites and other eruptive rocks; while traces of hydrocarbons, sulphur dioxide, oxygen, and nitrogen have also been noticed.

(3) *Fluid cavities* occur in crystals of topaz, corundum, beryl, diamond, and especially in quartz. The shape of the cavity is generally irregular, but in some cases it is the same as that of the host (negative crystal). The fluid is most commonly water or an aqueous

solution; sometimes it is liquid carbon dioxide. A bubble of air or of some gas is generally present which is frequently movable. Small crystals also float in some of the liquids, whose size increases and diminishes with changes of temperature. Crystals which grow rapidly from aqueous solutions very frequently imprison portions of their mother-liquor.

(4) *Glass inclusions* occur in crystals formed from a molten mass. They are inclusions of the mother-liquor which have solidified in an amorphous state. They also often contain one or more bubbles, surrounded by wide black rims, and of course immovable. The presence of more than one bubble in one cavity is indicative of a glass incision.

(5) *Inclusions of unindividualized matter* are for the most part quite irregular, like clay in rock-salt, gypsum and quartz; bituminous substances in andalusite and quartz; iron hydroxide in mica and gypsum.*

(6) *Minute crystals* of various kinds are often included in larger individuals. These may impart a peculiar color or lustre to the latter, as in the case of the hematite plates in carnallite and oligoclase (sunstone). The microscope has done much to extend our knowledge of such associations, which are very manifold. The iridescence of minerals like labradorite, hypersthene and bronzite is due to minute inclusions of other minerals arranged in certain crystallographic planes. These are regarded by Professor Judd as of secondary origin, occupying cavities produced in planes of easiest solution by percolating waters at con-

---

* See Blum, Leonhard, Seyffert, and Söchting: Die Einschlüsse von Mineralien. Haarlem, 1854.

siderable depths. Professor Judd has termed this process *schillerization*.

In other cases quite similar results are produced by original inclusions, as in the case of the sanidine of some of the recent rhyolites, and the zonally arranged microlites of hauyne and many of the Mt. Somma minerals.

In still other cases, minute mineral impurities may be developed by the incipient alteration of their host. By this means many crystals of feldspar contain flakes of muscovite or kaolin, as well as needles of zeolites, scapolite, or zoisite. This subject is very extensive, and leads directly into the field of pseudomorphism, metamorphism and chemical geology, which lie without the scope of the present work.

APPENDIX.

ON ZONES, PROJECTION AND THE CONSTRUCTION OF CRYSTAL FIGURES.

It is impossible to include within the limited space of the present work the explanations and formulæ necessary for the mathematical calculation of crystallographic constants and symbols from observed interfacial angles. The simple zonal relations existing between crystal planes and their graphic representation by means of projections are, however, easily comprehended and are therefore useful, even to the beginner. While these subjects are not absolutely essential to the understanding of what has been given in the body of this book, they may nevertheless be employed with advantage in connection with the study of each system. Their consideration is therefore embodied in an appendix which may be employed as desired.

Zones.

Definition. A study of the relation existing between the planes occurring on crystals is much facilitated by the fact that they are frequently arranged in belts, which extend around the crystal in different directions. Such belts of planes are technically called *zones.* All planes belonging to the same zone are said to be *tautozonal.* The intersection-edges of all tautozonal planes are parallel to each other; and, like the planes themselves, are also parallel to an imaginary line passing through the centre of the crystal, called the *zonal axis.* The positions or symbols of any two planes belonging to a zone are sufficient to determine the direction of its zonal axis. The symbol of any crystal plane is known if it be found to lie at the same time in two zones the directions of whose zonal axes are known.

The real nature of a zone may be made clearer by an example. In Fig. 354 the planes d, b, g, c, are tau-

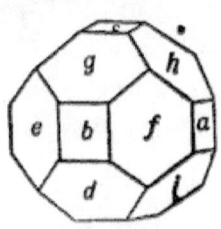

Fig. 354.

tozonal; likewise the planes d, f, h; e, b, f, a; i, f, g; etc. Here we have several distinct zones existing on the same crystal; while the plane b belongs equally to the two zones $d\ b\ g\ c$ and $e\ b\ f\ a$.

In Fig. 355 the planes oao' form a zone; and it is equally evident that the planes $a\ d\ d'$ also form one, in spite of the fact that d and d' do not actually intersect. If these two planes were extended until they did intersect, their edge would be parallel to that between a and d.

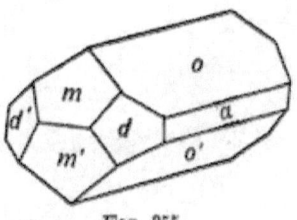

Fig. 355.

The practical determination of what planes belong to the same zone is accomplished, so far as possible, by the parallelism of their intersecting edges. In cases where there is any doubt as to the exact parallelism, or where the planes in question do not intersect, recourse is had to the reflecting goniometer (p. 22). It will be readily understood that if a zonal axis be made coincident with the axis of revolution of such an instrument, then each face of the zone must in turn yield a reflection as the crystal is revolved through 360°. This is the most common and ready method of identifying the planes which compose a crystal zone.

General Expression for the Indices of a Zone. The essential feature of any zone is the *direction* of its intersection-edges, or of its zonal axis. The indices of this direction may be obtained from the indices of any two planes lying in the zone, and they are called the *indices of the zone*.

It is capable of geometrical demonstration that, if the indices of any two planes belonging to a zone be indicated by the letters hkl and $h'k'l'$, then the indices of their zonal axis, which are usually designated by the letters $u, v, w,$ are equal to

$$(kl' - k'l)a, \quad (lh' - l'h)b, \quad \text{and} \quad (hk' - h'k)c.*$$

The zonal indices may be derived from the indices of any two planes in the zone by the following easily remembered process: The indices of the first plane are written twice in their usual order, and those of the second plane are placed directly under them in the same order. The first and last terms are then cut off. The product of the first upper and second lower indices has then subtracted from it the product of the second

---

* The complete proof of this, which is very simple, is too long to be given here. It will be found in Groth's Physikalische Krystallographie, 2d ed., p. 200 (1885), and in Bauerman's Systematic Mineralogy, p. 24 (1884).

upper and first lower indices. This gives u. v and w are similarly obtained, thus:

h	k	l	h	k	l
		×	×	×	
h'	k'	l'	h'	k'	l'

$(kl' - k'l) = u;\ (lh' - l'h) = v;\ (hk' - h'k) = w.$

It is very necessary, when using this formula, to pay particular attention to the *signs*.

We may illustrate this by the following concrete example: Suppose the indices of two planes are $0\bar{1}1$ and $\bar{2}\bar{1}1$, then the indices of their intersection or zonal axis will be $0\bar{2}\bar{2}$. These are obtained as follows:

0	$\bar{1}$	1	0	$\bar{1}$	1
		×	×	×	
$\bar{2}$	$\bar{1}$	1	$\bar{2}$	$\bar{1}$	1

$(\bar{1} \times 1) - (1 \times \bar{1}) = -1 - (-1) = 0 = u;$
$(1 \times \bar{2}) - (0 \times 1) = -2 - 0 = \bar{2} = v;$
$(0 \times \bar{1}) - (\bar{1} \times \bar{2}) = 0 - 2 = \bar{2} = w.$

The *algebraic* sum, remainder, and product must in every case be used, or an erroneous result will be obtained.

Zone Control. The indices of any plane which belongs to a zone whose zonal indices are known, must, when multiplied with the latter, give an algebraic sum equal to zero. If p, q, r, are the indices of the plane in question, then

$$up + vq + wr = 0.*$$

---

* The proof of this equation is again too long to be quoted here. It will be found in Groth's Physikalische Krystallographie, 2d ed., p. 204 (1885).

APPENDIX.

This is called the *zonal equation*, and is used as a control in deciding whether a plane belongs to a given zone or not. We may apply it to test the result of our last example. If the planes $0\bar{1}1$ and $\bar{2}\bar{1}1$ really belong to the zone whose indices are $0\bar{2}\bar{2}$, then

and
$$(0 \times 0) + (\bar{2} \times \bar{1}) + (\bar{2} \times 1) = 0$$
$$(0 \times \bar{2}) + (\bar{2} \times \bar{1}) + (\bar{2} \times 1) = 0.$$

The zonal equation may also be sometimes used to determine the symbol of a plane which truncates an edge between two known planes and therefore lies in a zone with them. For example, the edges of the rhombic dodecahedron, ∞O, are truncated by the faces of a form mOm (Plate II., Fig. 18); what is the value of m? The edge between the planes whose indices are 101 and 011 is replaced by a face whose intercepts are $ma : ma : a$, and whose indices are consequently $1, 1, m$. The indices of the zonal axis of the two known planes are found, as explained in the last paragraph, to be $\bar{1}\bar{1}1$; hence

$$\bar{1} + \bar{1} + (1 \times m) = 0; \quad \text{and} \quad m = 2.$$

The Integrity of Zones. From the last section it will be seen that the existence of zones is wholly dependent on the indices of the planes which compose them, and quite independent of the relative lengths of the axes to which these are referred. These latter values are found to vary slightly with the temperature, since the expansion of a crystal by heat is unequal in different directions. The indices, however, retain their rational values for all temperatures, and therefore no change of external conditions can affect the so-called *integrity of the zones*.

Determination of the Indices of an Unknown Plane belonging to Two Known Zones. Any plane which is parallel to two different and known directions has its position thereby determined. A crystal plane which lies simultaneously in two zones must be parallel to the zonal axes of both. If the indices of two zones are u, v, w, and u', v', w', and the indices of a plane belonging to both, p, q, r, then

$$up + vq + wr = 0$$

and

$$u'p + v'q + w'r = 0.$$

From these two zone-control equations we obtain

$$p = r \cdot \frac{vw' - v'w}{uv' - u'v}$$

and

$$q = r \cdot \frac{wu' - w'u}{uv - u'v'}.$$

Inasmuch as it is possible to make one of the indices equal to any number without destroying their relative values, we may make $r = uv' - u'v$ and thus obtain

$$p = vw' - v'w;$$
$$q = wu' - w'u;$$
$$r = uv' - u'v.$$

These are the values of the indices of the plane expressed in terms of the zonal indices.

The indices of an unknown plane, found to lie simultaneously in two known zones, may therefore be found by combining the zonal indices according to the same

APPENDIX. 223

method as was above given for finding the zonal indices from the indices of two known planes. Thus:

$$\frac{\begin{array}{c|cccc|c} u & v & w & u & v & w \\ & \times & \times & \times & & \\ u' & v' & w' & u' & v' & w' \end{array}}{vw' - v'w = p; \quad wu' - w'u = q; \quad uv' - u'v = r.}$$

Example. Suppose, on a certain crystal, an unknown plane is found to lie in one zone with two other planes whose indices are 110 and 101; and at the same time to belong to a second zone two of whose planes have the symbols 111 and 100. What are the indices of the plane in question?

$$\frac{\begin{array}{c|cccc|c} 1 & 1 & 0 & 1 & 1 & 0 \\ & \times & \times & \times & & \\ 1 & 0 & 1 & 1 & 0 & 1 \end{array}}{(1-0)\ (0-1)\ (0-1)}$$

$(1\bar{1}\bar{1})$, indices of 1st zone.

$$\frac{\begin{array}{c|cccc|c} 1 & 1 & 1 & 1 & 1 & 1 \\ & \times & \times & \times & & \\ 1 & 0 & 0 & 1 & 0 & 0 \end{array}}{(0-0)\ (1-0)\ (0-1)}$$

$(01\bar{1})$, indices of 2d zone.

$$\frac{\begin{array}{c|cccc|c} 1 & \bar{1} & \bar{1} & 1 & \bar{1} & \bar{1} \\ & \times & \times & \times & & \\ 0 & 1 & \bar{1} & 0 & 1 & \bar{1} \end{array}}{(1-\bar{1})\ (0-\bar{1})\ (1-0)}$$

(211), indices of plane required.

PROJECTION.

Definition. The zonal relations of crystal planes are advantageously represented by graphic methods. Two such methods are at present extensively employed, both of them being systems of *projection* proposed by Neumann in 1823.

The first is called *spherical projection*, and represents

the position of each plane on the upper half of a crystal by the point of intersection of a normal to the plane with the surface of a sphere at whose centre the crystal is supposed to be. The hemisphere thus obtained is then projected on a plane passed through its equator, while the eye is imagined at the opposite extremity of its polar diameter.

The second method is called the *linear projection*. It represents each crystal face by its line of intersection with an imaginary plane, called the *plane of projection*. While neither so elegant nor so useful for purposes of calculation as the first method, this projection possesses certain peculiar advantages for beginners, and is the basis upon which crystal drawings are constructed. For these reasons it will be first considered.*

Construction of Linear Projections. If we imagine all the faces of a crystal to be shifted without changing their direction, until they cut the vertical axis at unit distance from the centre, then their linear projection is formed by the lines of their intersections with an imaginary plane (plane of projection) which passes through the lateral axes. Any plane may be made the plane of projection, but the one here mentioned is generally selected for this purpose. The lines of intersection between the crystal faces and the plane of projection are called the *section-lines* of the planes. Every section-line stands for a pair of planes on all holohedral and parallel-face hemihedral forms; but projec-

---

* Those desiring full information regarding the construction and use of spherical projections will find it in the works of Miller, Reusch, Liebisch, Groth, and Henrich, cited at the beginning of this book. The works of Quendstedt, Klein, and Websky contain similar details in reference to linear projections.

tions of inclined-face forms do not differ from those of the corresponding holohedrons, because the section-lines here stand only for single faces.

The construction may be advantageously illustrated by a few simple examples.

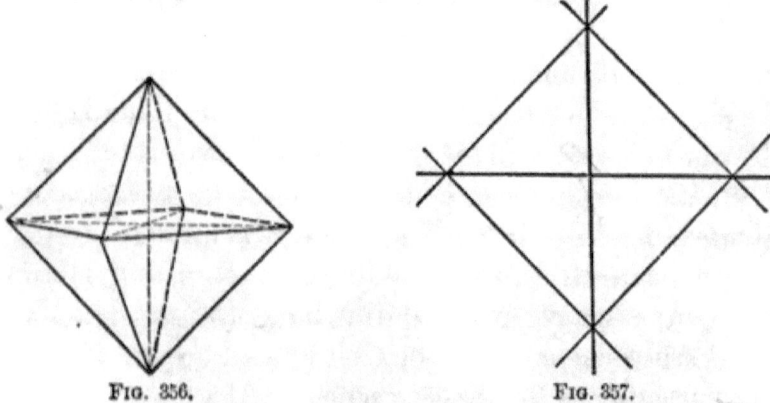

Fig. 356. Fig. 357.

The Octahedron. The planes of this form already cut the vertical axis at unity, and hence require no shifting. The intersection of the four upper planes with the horizontal axial plane would evidently give the projection in Fig. 357.

The Cube. The upper surface of the cube evidently cannot appear in the projection since it cuts the ver-

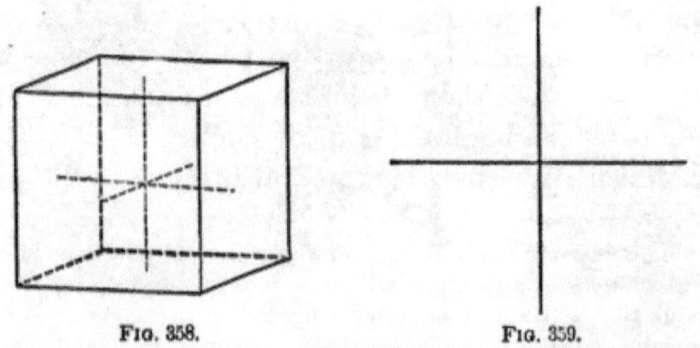

Fig. 358. Fig. 359.

tical axis at unity, but intersects the plane of projection only at infinity (i.e. is parallel to it). The other planes, in order to be made to cut the vertical axis at unity, must be shifted until they include it throughout. Each plane will then coincide with its parallel plane, and the two pairs will intersect the plane of projection in a rectangular cross, Fig. 359.

The most simple rule for constructing a linear projection is to lay off on paper (the plane of projection) the two (or three) lateral axes which lie in this plane. Then reduce the Weiss symbol for each plane in the upper half of the crystal to a form in which its vertical axis is equal to unity. [This is, of course, accomplished by dividing each term of the symbol by the parameter of the vertical axis.] When the symbols have all been reduced to this form, the section-line for any plane may be obtained by merely connecting those points on the lateral axes, whose positions are indicated by the relative values of the new lateral parameters.

Examples. The form shown in Fig. 360 is the isometric icositetrahedron (p. 55), composed of three planes in each octant whose symbols, according to Weiss's notation, are:

$$\left\{ \begin{array}{l} 2a : 2a : a \\ 2a : a : 2a \\ a : 2a : 2a \end{array} \right\}.$$

These symbols, when reduced to a form in which $c = 1$, become:

$$\left\{ \begin{array}{l} 2a : 2a : a \\ a : \tfrac{1}{2}a : a \\ \tfrac{1}{2}a : a : a \end{array} \right\}.$$

From these values the projection shown in Fig. 361 is readily constructed.

APPENDIX.

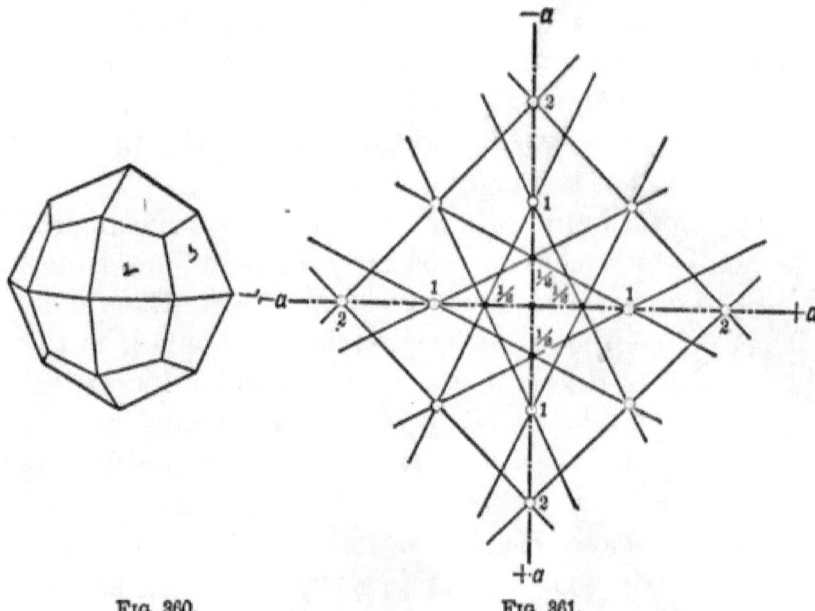

Fig. 360.   Fig. 361.

The next example represents a rhombohedral combination occurring on the mineral tourmaline (Fig. 362). Its eight forms are as follows:

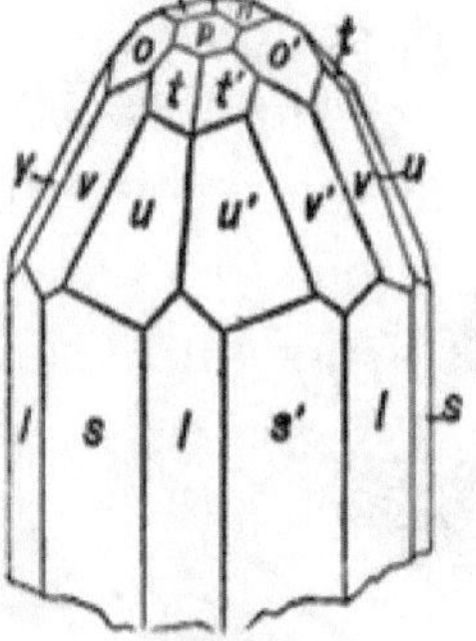

∞P, $\{10\bar{1}0\}$ (l); $\infty P2$, $\{11\bar{2}0\}$ (s);
R, $\kappa\{10\bar{1}1\}$ (p); $-\frac{1}{2}R$, $\kappa\{01\bar{1}2\}$ (n);
$-2R$, $\kappa\{02\bar{2}1\}$ (o); R^3, $\kappa\{3\bar{1}21\}$ (t);
R^5, $\kappa\{5\bar{2}31\}$ (u); $-2R^2$, $\kappa\{13\bar{4}1\}$ (v).

Fig. 362.

The zonal relations of these planes are manifold, and may be exhibited with great distinctness upon a linear projection like that shown in Fig. 363.

The third example shows a triclinic crystal whose planes are referred to axes of unequal length and obliquely inclined to each other. The two lateral axes intersect at an angle of 131° 33′. The Weiss symbols of the planes are as follows:

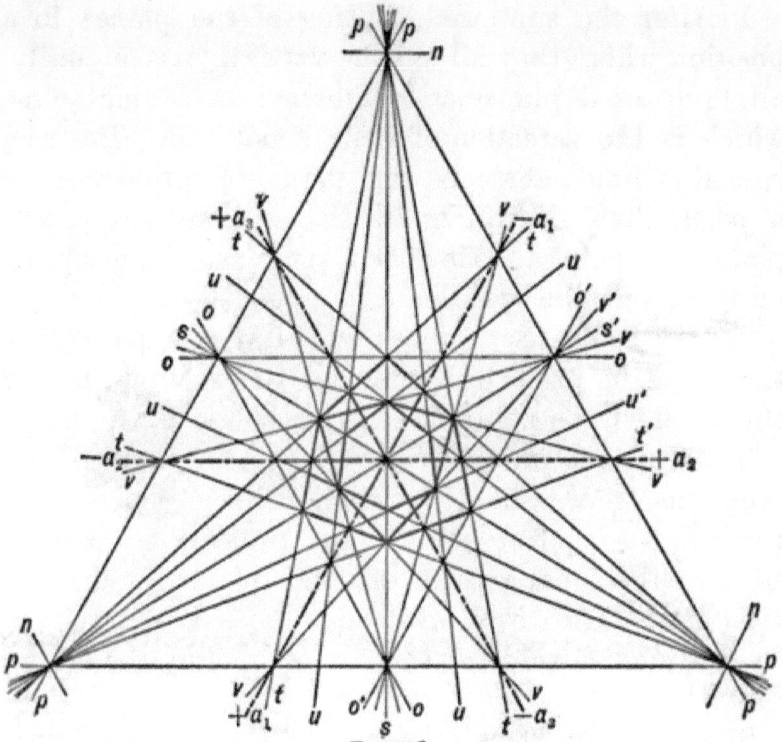

228   APPENDIX.

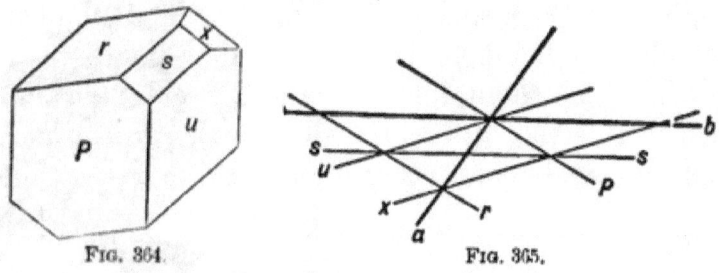

Fig. 363.

$$\begin{cases} P = a : -b : \infty c \\ u = a : b : \infty c \end{cases} \qquad \begin{cases} r = a : -b : c \\ x = a : b : c \\ s = a : \infty b : 2c \end{cases}$$

Axial ratio: $a : b = 0.492 + : 1$.

Fig. 364.   Fig. 365.

A careful consideration of these examples, together with what has been said before with reference to the linear projection, will make clear the truth of the following important points:

1. After the supposed shifting of the planes to a position where they all cut the vertical axis at unity, all tautozonal planes will intersect in a single line which is the direction of their zonal axis. But as a rule this line intersects the plane of projection in a *point*, through which all the section-lines of the planes belonging to this zone must pass. Such a point on the projection is called a *zone-point*.

2. If the direction of any zonal axis is parallel to the plane of projection, then all the section-lines of this zone will be parallel, i.e. will intersect at infinity.

3. All planes which are parallel to the vertical axis must be represented by section-lines which pass through the central point of the projection. This is because the zonal axis for all such planes is the vertical axis. The *direction* of such lines is determined by the relative value of their intercepts on the lateral axes.*

Symbol of any Plane belonging to Two Zones obtained by Linear Projection. The linear projection presents another ready means of obtaining the symbol of any plane lying simultaneously in two zones. The intersection of any two section-lines in the projection is enough to fix the position of the zone-point of the zone to which they belong. If the two zone-points of the zones in which the unknown plane lies can be determined in this way, its section-line in the projection is found by merely connecting them. From the section-

---

* The construction of linear projections of various crystal forms should be made a matter of constant practice by the student until the subject presents no difficulty, and until he can appreciate the full significance of such projections at a glance.

230   APPENDIX.

line thus obtained the symbol of the plane can readily be deduced.

Example. Suppose a plane lies in a zone with the planes $a : b : \infty c$ and $a : \infty b : c$, and at the same time in a second zone with the planes $a : b : c$ and $a : \infty b : \infty c$; what is its symbol?

The symbol of the plane (dotted section-line in Fig. 366) must be: $\frac{1}{2}a : b : c = a : 2b : 2c$. This is the same case as that solved by the other method (p. 223), and the results will be seen to agree.*

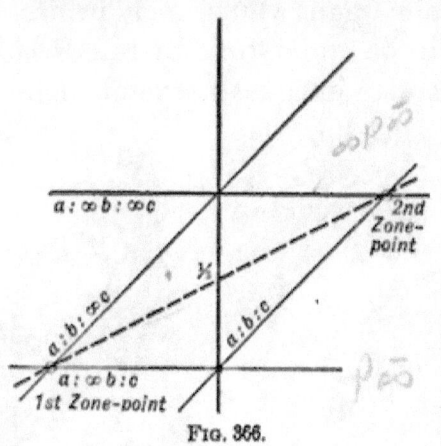

Fig. 366.

---

* A few examples for practice are here given. After they have been solved by both of the above-explained methods other examples should be taken until the subject is satisfactorily mastered.

What plane lies in each of the following pairs of zones?

	Weiss's Symbols.		Naumann's Symbols.	Miller's Symbols (Indices).	
1.	$a : \infty a : \infty a$	$\infty a : a : \infty a$	$\infty O \infty : \infty O \infty$	100	010 (1st zone)
	$a : 2a : 2a$	$a : 2a : -2a$	$2O2 : 2O2$	211	$21\bar{1}$ (2d zone)
2.	$\infty a : \infty a : a$	$a : 2a : \infty a$	$\infty O \infty : \infty O2$	001	210
	$a : \infty a : \infty a$	$a : a : a$	$\infty O \infty : O$	100	111
3.	$\infty a : \infty a : a$	$a : 2a : \infty a$	$\infty O \infty : \infty O2$	001	210
	$a : \infty a : \infty a$	$2a : a : 2a$	$\infty O \infty : 2O2$	100	121
4.	$a : -a : \infty c$	$a : a : c$	$\infty P : P$	$1\bar{1}0$	111
	$a : -a : 2c$	$a : a : \infty c$	$2P : \infty P$	$2\bar{2}1$	110
5.	$a : a : 3c$	$a : -a : 3c$	$3P : 3P$	331	$3\bar{3}1$
	$a : -a : \infty c$	$a : a : c$	$\infty P : P$	$1\bar{1}0$	111
6.	$a_1 : \infty a_2 : -a_3 : 2c$ $2a_1 : 2a_2 : -a_3 : \infty c$		$2P : \infty P2$	$20\bar{2}1$	$11\bar{2}0$
	$2a_1 : 2a_2 : -a_3 : 2c$ $a_1 : \infty a_2 : -a_3 : \infty c$		$2P2 : \infty P$	$11\bar{2}1$	$10\bar{1}0$

APPENDIX. 231

Spherical Projection. The other method of graphically representing the relationships of crystal planes mentioned on p. 224 is called *spherical* or *stereographic* projection. The principle upon which such projections are constructed may be understood by reference to Fig. 367. Suppose that we imagine a crystal—here

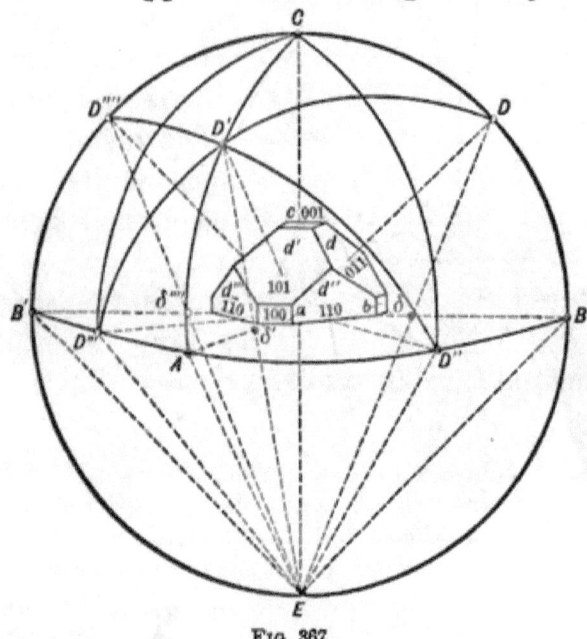

FIG. 367.

the simple isometric combination of cube and rhombic dodecahedron—so placed that its center coincides with the center of a sphere, while its vertical axis is also coincident with the vertical axis of the sphere. If lines be drawn from the center of the sphere, normal to each of the crystal planes, they will intersect the surface of the sphere in points, called the *poles* of the planes. The distribution of the poles A, B, B', C, D, D', etc., on the upper hemisphere, definitely fixes the relative positions of the planes a, b, b', c, d, d', etc., on the crystal.

The arrangement of the poles on the upper hemisphere is represented on a plane surface as it would appear to the eye situated at the lower extremity of the vertical axis, E; in other words, the poles are *projected* on the equatorial circle of the sphere, $BD''AD'''B'$, which is called the *fundamental circle*, (German, *Grundkreis*). In this way the projection of any pole becomes the point of intersection between a line joining it with the lower end of the meridian axis, E, and the equatorial circle. Thus the projection of the pole D is δ; of D', δ'; etc. The complete spherical projection of the combination shown in Fig. 367 is given in Fig. 368.

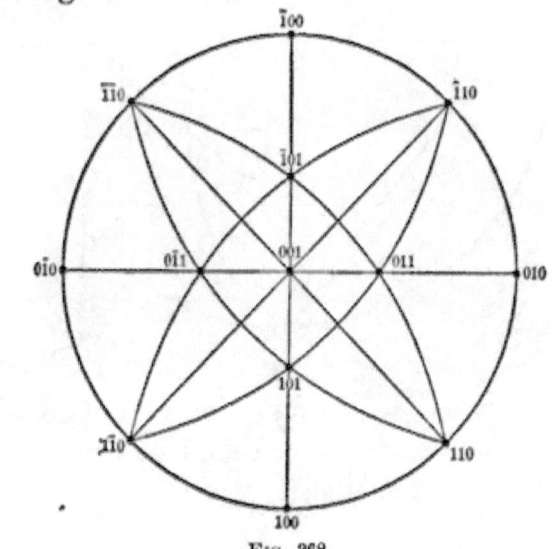

Fig. 368.

A consideration of the above-described example will illustrate the following properties of spherical projections:

1. For all crystals with rectangular axes the pole of the basal pinacoid {001} will occupy the central point of the projection.

2. The poles of all planes belonging to the prismatic zone, i.e. parallel to the vertical axis, will lie in the circumference of the fundamental circle.

3. The poles of all planes belonging to the same zone will fall in the circumference of the same great circle (*zonal circle*); and the same is true of the projections of such poles.

4. All zonal circles whose axes are horizontal appear in the projection as diameters of the fundamental circle.

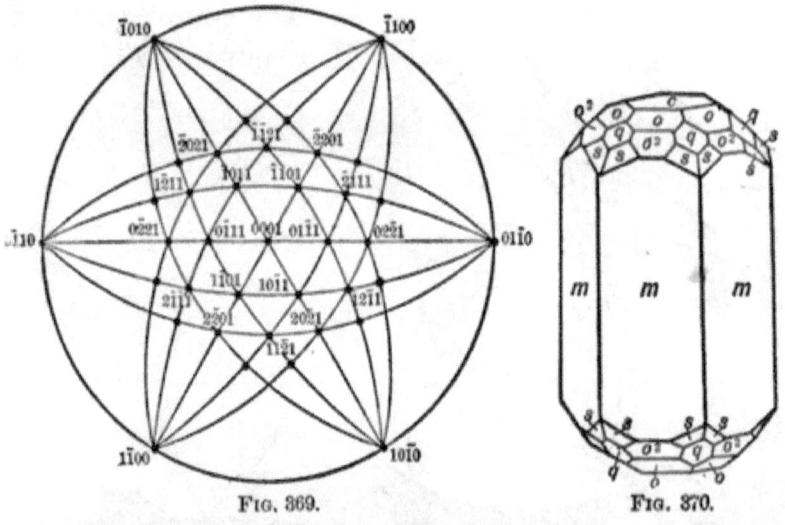

Fig. 369.   Fig. 370.

5. The angular distance between any two poles or their projections, measured on their zonal circle, is equal to the *normal angle,* or supplement of the interfacial angle included between the planes to which the poles belong.

The last-named property of spherical projections renders them particularly valuable as aids in crystallographic calculation, but the details of their application to this end, as well as the method of their practical construction, must be sought for in larger works.

APPENDIX.

The spherical projection of a crystal admirably expresses its symmetry, as may be seen from the two following examples. Fig. 369 is the projection of the holohedral hexagonal combination observed on beryl (Fig. 370) as described on page 116. The dots without indices are the projections of the poles of the twelve faces of the dihexagonal pyramid, $3P\frac{3}{2}$, $\{32\bar{1}1\}$ (*s*).

Fig. 371 shows by spherical projection the relation of the planes which commonly occur on the monoclinic feldspar, orthoclase (Fig. 372), described on page 166.

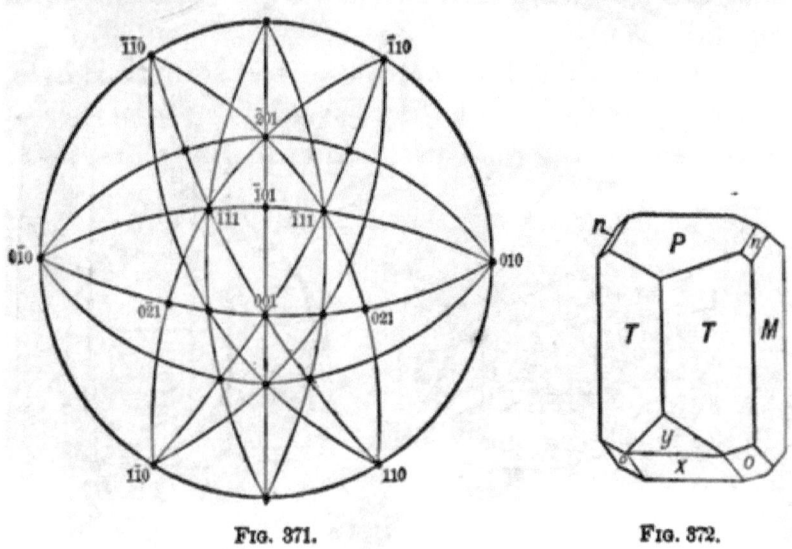

FIG. 371.   FIG. 372.

THE CONSTRUCTION OF CRYSTAL FIGURES.

Method of Representing Crystal Forms. The accurate construction of crystal figures is a matter of much importance, particularly in connection with descriptions of crystals of new substances or of such as possess unusual or complicated habits. It is customary to represent crystals in their ideal development, i.e. as free from all distortion of form or irregularity of growth, unless these possess some peculiar significance.

The parallelism of crystal edges is so important, as indicating the existence of zones, that it is most desirable to retain this feature in the crystal figure. For this reason ordinary perspective figures are not employed, but rather such as represent the crystal at an infinite distance from the observer. In this way all rays coming to the eye are parallel, and a projection is formed wherein all the edges belonging to the same zone are parallel in the figure, as they are on the actual crystal.

Projections like those here described are of two kinds, according as the parallel rays passing from the crystal to the eye are normal or oblique to the plane upon

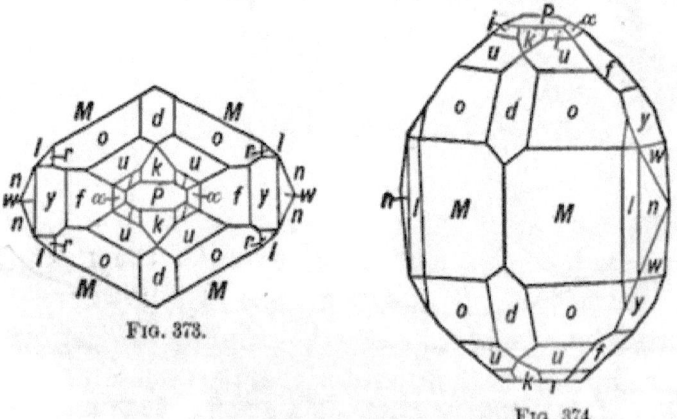

Fig. 373.

Fig. 374.

which the figure is projected. The former are called *orthographic*, and the latter *clinographic* projections.

Orthographic Projections. It is customary to construct orthographic projections of crystals upon a horizontal plane (which in all systems with rectangular axes is the basal pinacoid), while the eye is conceived of as at an infinite distance in the direction of the vertical axis.

Thus, Fig. 373 represents the orthographic projection upon the basal plane of a complicated crystal of topaz, whose vertical clinographic projection is given in Fig. 374. The forms of this combination are OP, $\{001\}$ (P); ∞P, $\{110\}$ (M); $\infty P\breve{2}$, $\{120\}$ (l); $\infty P\bar{4}$, $\{140\}$ (n); P, $\{111\}$ (o); $\frac{1}{2}P$, $\{112\}$ (u); $\frac{1}{3}P$, $\{113\}$ (i); $P\breve{\infty}$, $\{101\}$ (d); $\frac{1}{3}P\breve{\infty}$, $\{103\}$ (k); $\frac{2}{3}P\breve{\infty}$, $\{023\}$ (a); $P\bar{\infty}$, $\{011\}$ (f); $2P\breve{\infty}$, $\{021\}$ (y); $4P\breve{\infty}$, $\{041\}$ (w); and $2P\breve{2}$, $\{121\}$ (r).

The most ready and convenient way to construct an orthographic projection of any crystal is to prepare a linear projection (p. 224) of all of its forms upon the same plane as that selected for the orthographic projection. If now the central point of this linear projection is connected with the point of intersection of any two section-lines, the direction of the edge between the two planes corresponding to these section-lines is thereby obtained for the orthographic projection.

It is sometimes desirable to construct orthographic projections of monoclinic crystals on the plane of their clinopinacoid. When either this or the basal pinacoid is selected as the plane of projection the method of procedure is that above given. If, however, it is desired to construct an orthographic projection of a monoclinic crystal upon a plane perpendicular to the vertical axis, the clinodiagonal axis should not then be represented at its full length as compared with the orthodiagonal, but its length should be $\dot{a} \cdot \sin \beta$.

If it is desired to construct an orthographic projection of a triclinic crystal upon a plane perpendicular to the vertical axis, the two lateral axes should be made to intersect at the angle included between its two vertical pinacoids, while their lengths should be proportional to the values $\breve{a} \cdot \sin \beta$ and $\bar{b} \cdot \sin \alpha$.

These values must be employed in making the linear projection, from which the orthographic projection is derived as above described.

Clinographic Projections. If the eye, still conceived of as at an infinite distance, is not directly in front of the crystal, but to one side, then the parallel rays which reach it are oblique to the plane of projection and a clinographic projection results.

It is customary to represent crystals by their clinographic projections drawn in a vertical position with the eye turned a certain angular distance (δ) to the right, and elevated a certain angle (ϵ) above the center of the crystal. In this way its right side and top are brought into view. Such projections differ from ordinary perspective figures in having no vanishing point, i.e. they show as parallel all lines, whatever is their direction, which are parallel on the object. They are therefore examples of parallel perspective.

In order to construct a clinographic projection of a crystal, it is necessary to know (1) the values of its crystallographic constants, and (2) the crystallographic symbols of its planes. The first step in the construction of the projection is then the preparation of a perspective view of the axes, in exactly the position desired for the finished figure.

It is usual to assume such values for the angular revolution and elevation of the eye (δ and ϵ) as can be expressed by a simple ratio between the projected axes, when their actual lengths are equal (isometric system).

The values of the angles δ and ϵ are determined as $\cot \delta = r$, $\cot \epsilon = rs$, it being usual to make $r = 3$ and $s = 2$. In this case we have (Fig. 375)

$OI : OK' :: 1 : 3$ and $IA : IO :: 1 : 2$,

when the value of δ is 18° 26′, and of ϵ, 9° 28′.

Projection of the Axes for the Isometric System. If the values for r and s be assumed as above, then the method of construction of the isometric axes is as follows (Fig. 375): Draw two lines LL' and KK' at right angles to one another. Make $KO = K'O$, and divide KK' into three equal parts. Draw verticals through the four points thus obtained on KK', and below K' lay off $K'H = \tfrac{1}{2}K'O$. Draw HO, which will give the direction of the front lateral axis. Its length will be that portion of this line included between the two inner verticals, A and A'.

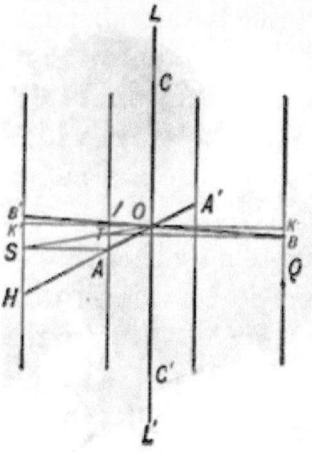

Fig. 375.

Draw AS parallel to $K'O$ and connect the points S and O. From the intersection of this line with the inner vertical, T, draw TB parallel to $K'K$. From point, B, thus obtained draw the line BB' through O. This will be the second lateral axis.

Below K, lay off $KQ = \tfrac{1}{3}OK$ and make $OC = OC' = OQ$; then CC' will be the length of the vertical axis.

Projection of the Axes for the Tetragonal and Orthorhombic Systems. The axes constructed for the isometric system may be readily adapted to both the other systems with rectangular axes by merely laying off portions of the lines AA' and CC' (Fig. 375), which are proportional to the lengths expressed in the axial ratios of the crystals to be figured.

In the case of a tetragonal crystal like zircon, whose axial ratio is $a : c :: 1 : .64$, the two lateral axes remain unchanged, while the vertical axis must be made .64 of the length CC'.

For an orthorhombic crystal the axis BB' alone remains unchanged, while AA' and CC' are both reduced to the proportionate lengths belonging to the substance in question.

Projection of the Monoclinic Axes. To project the inclination, β, of the clinodiagonal axis, we construct the axes as in the isometric system, and then lay off $Oc = OC \cdot \cos \beta$, and on OA' lay off $Oa = OA' \cdot \sin \beta$. From c draw a line parallel to OA', and from a another parallel to OC. From their intersection, a line (DD') drawn through O will give the direction of the clino-axis (Fig. 376). The relative lengths of the axes must now be determined, according to the axial ratio of the substance, as in the orthorhombic system.

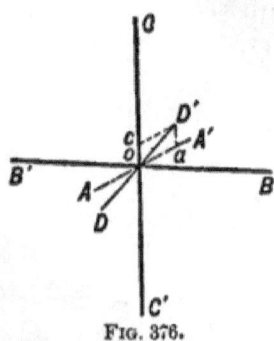

Fig. 376.

Projection of the Triclinic Axes. In this case all three axes of reference intersect obliquely $b \wedge c = \alpha, a \wedge c = \beta, a \wedge b = \gamma$. If we start with the isometric axes, the first step in their adaptation to the triclinic system is to obtain the direction of the two vertical axial planes or pinacoids. To do this, we lay off on OB, $Ob = OB \cdot \sin \phi$ (ϕ being the angle $\infty P \bar{\infty}$, $\{100\} \wedge \infty P \breve{\infty}$, $\{010\}$, which is evidently not the same as γ), and on OA, $Oa = OA \cdot \cos \phi$ (Fig. 377). The line drawn from the angle d of the parallelogram $adbO$ through O will give the direction of the macropina-

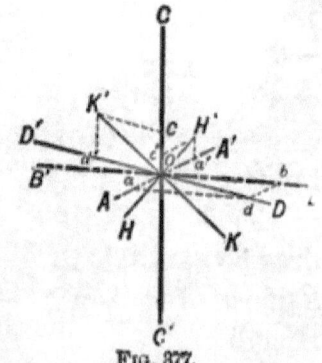

Fig. 377.

coidal section, DD'. To obtain the direction of the macrodiagonal axis ($\bar{b}$), lay off on OD', $Od' = OD^{\bullet} \cdot \sin \alpha$; and on OC, $Oc = OC \cdot \cos \alpha$. From the parallelogram, $d'OcK'$, thus obtained, the diagonal, $K'K$, gives the macrodiagonal axis. In a similar manner, the brachydiagonal axis ($\breve{a}$), HH', is found by laying off on OA', $Oa' = OA' \cdot \sin \beta$; and upon OC, $Oc' = OC \cdot \cos \beta$. After the axes have been projected in their proper directions, their relative lengths must be given them in accordance with the axial ratio of the substance, just as in the orthorhombic and monoclinic systems.

Projection of the Hexagonal Axes. These may be projected in a manner analogous to that given for the isometric axes, as explained by Dana (see literature references below). A simpler method is, however, as follows: Construct an orthorhombic set of axes whose axial ratio, $\breve{a} : \bar{b} : \breve{c}$, is

$$\sqrt{3} \, (= 1.732) : 1 : \breve{c}$$

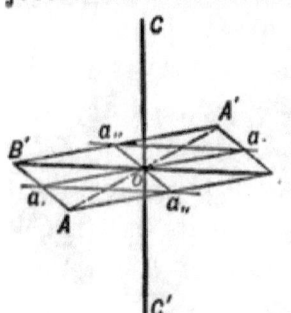

Fig. 378.

($\breve{c}$ being given the value of the vertical axis belonging to the substance to be drawn); connect the extremities of the two lateral axes, and, in the rhomb thus formed, the obtuse angles, at the ends of the b axis, will be exactly 120°. If lines be now drawn parallel to $\bar{b}$, through points on the axis, $\breve{a}$, half way between its extremities and the center, o, the rhomb will be converted into a hexagon, with all of its angles exactly 120°. If we connect the diagonally opposite angles of this hexagon, we shall obtain the projection of the hexagonal axes required (Fig. 378).

Construction of Crystal Figures upon the Axes. After the axes for any particular substance have been constructed according to the methods above explained, the next step is to erect upon them the complete clinographic projection of the crystal whose figure is desired. The manner in which this is accomplished is the same for all the systems. Such figures consist of a series of lines representing in parallel perspective the combination-edges between the crystal planes. It is first necessary to determine the proper *direction* of each of these edges, and then they may be united so as to represent the particular combination or habit desired. For the latter purpose it is desirable to have as a guide a free-hand sketch showing approximately the relative development and distribution of the various planes.

The simpler forms of each system may be constructed directly upon the projected axes in a way which requires no particular explanation. For instance, connecting the extremities of the axes produces the ground-form of the substance to be drawn. Vertical lines through the extremities of the lateral axes give the fundamental prism, etc.

This method is, however, not readily applicable to the construction of figures of more complex combinations. Another method is therefore usually employed for these, which is based upon the use of the linear projection. A complete linear projection of the crystal to be drawn is first prepared in the ordinary manner (p. 224), and this is then thrown into parallel perspective upon the axes by connecting points on the lateral axes, which correspond to those so connected on the linear projection. When this has been done, *the direction of any combination-edge is found by merely connecting the point where the two section-lines* (representing the planes forming the edge in question) *meet with the unit distance on the vertical axis.* The directions of the various edges thus obtained may be combined in the way they appear on the crystal. After lengths have been assumed for one or two of the lines, according to the size of the drawing desired, the lengths of the other lines will be determined by their mutual intersections.

This method of construction may be advantageously illustrated by an example. Required to project in parallel perspective a crystal of sulphur, showing the forms: $OP, \{001\}(c); P, \{111\}(p);$ $\frac{1}{3}P, \{113\}(s);$ and $P\infty,$ $\{011\}(n).$ The axial ratio of this substance is $0.813 : 1 : 1.904$. Fig. 379 gives an ordinary linear projection of these forms, and Fig. 380, the same thrown into perspective upon a set of

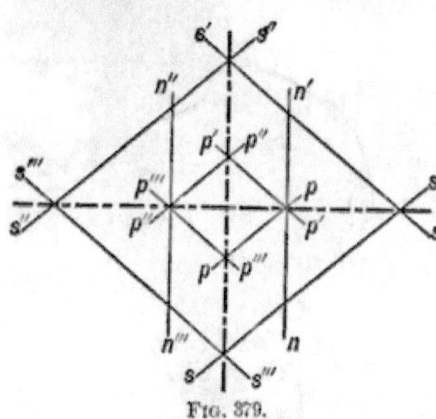

Fig. 379.

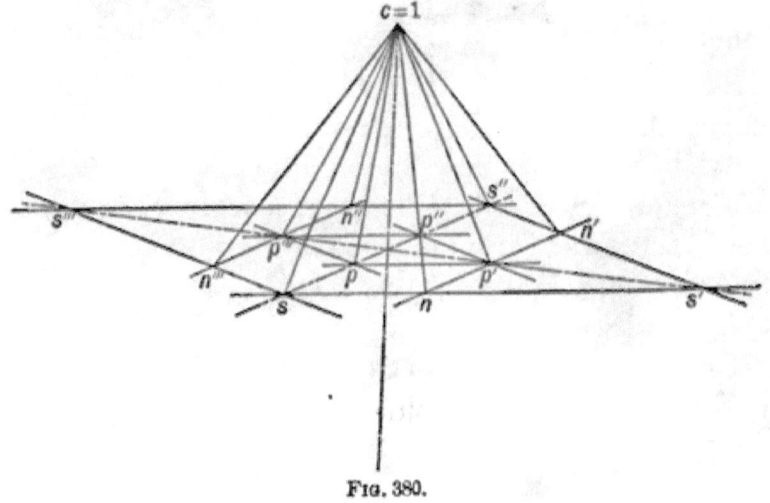

Fig. 380.

orthorhombic axes having the lengths required by the axial ratio of sulphur. The shape of the basal pinacoid, *c*, in the figure is given by its intersections with *s*, and it must therefore be that of the rhomb, $ss's''s'''$. The size of this rhomb must be determined according to the dimensions desired for the figure.

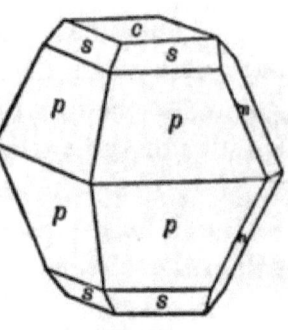

Fig. 381.

From the right-hand corner of the face, *c*, descend two lines representing the combination-edges of the plane *n* with the front and back, right-hand faces of the obtuse pyramid (*s* and *s'*). The directions of these two edges are obtained by joining the points *n* and *n'* (the intersections of the section-line of the plane *n* with those of the planes *s* and *s'*) with *c* ($=1$) (Fig. 380). Two lines with these directions should be drawn from the right-hand corner of the rhomb *c*, and the length of one of them determined so as to give to the face *s* approximately the relative size that it has on the crystal. When this is done, the length of the other line is determined by its intersection with a light line drawn from the lower extremity of the first one, parallel to the axis ss''.

In the same manner the direction of the combination-edge *s* : *s* is determined by connecting the point *s* with *c*; the edge *p* : *n* is obtained by joining *p'* and *c*; the edge *p* : *p* by connecting *p* and *c*; etc., etc.

It is of course necessary to determine the direction of only one edge in a zone, since all other edges belonging to this zone must be parallel to it. The recollection of this fact will greatly facilitate the construction.

The lower half of the crystal figure may be constructed either by joining the same points used in the upper half with the lower extremity of the vertical axis, or by using the posterior half of the perspective linear projection in connection with the upper extremity of the vertical axis.

After a certain amount of practice, complicated figures can be rapidly and accurately constructed on this principle. To avoid the confusion of too many lines in the perspective projection, the axes should be drawn in large proportions, and the directions of the edges not combined directly upon them, but on one side; the lines being transferred by a parallel ruler, or by a ruler and triangle, to another part of the paper. It is also desirable to draw the finished figure somewhat larger than the size intended for reproduction, and to have it subsequently reduced by photo-engraving. If the linear projection is very complex, it is not necessary to transfer it entire to the axes, but only such a portion of each section-line as, by its intersection with other lines, will give the zone-points required.

Construction of Figures of Twin Crystals. This depends entirely upon securing two sets of axes, one of which occupies a position as though it had been revolved 180° about a line normal to some given plane (the twinning plane). Suppose (Fig. 382) that abc represent the relative lengths of the axes, and XYZ the position of the twinning plane. It is required to construct a normal from O to the plane XYZ. From X draw XL parallel to ac; and from Z, ZH parallel to ac. Construct the parallelogram $OLDH$ and draw OD. In the same manner draw YL' and ZK, both parallel to bc, and construct the parallelogram $OL'FK$. Draw OF. If from the two points R and S straight lines be drawn to the opposite angles of the triangle XYZ,

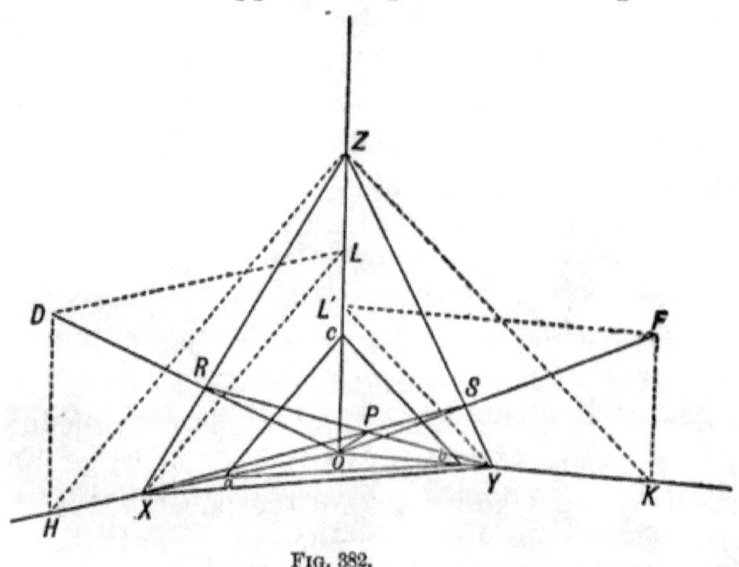

FIG. 382.

then their intersection, P, will be the point of emergence of a normal from O to the plane XYZ.

If now OP be prolonged to double its length at P' (Fig 383), and if this point be connected with X, Y and Z, then the lines $P'X$, $P'Y$, and $P'Z$ are the axes in the twinning position required, and of lengths corresponding to the parameters of the plane XYZ. To reduce them to unit lengths, corresponding to abc, we must draw aa', bb', and cc' all parallel to OP.

Upon this double set of axes the forms belonging to each individual are constructed in the same manner as has been explained for simple crystals. Each is drawn entire if penetration twins are to be represented, and in part if a figure of a contact twin is desired.

For the convenience of those desiring fuller information in regard to the construction of crystal figures, the following references are appended:

C. F. Naumann: Lehrbuch der reinen und angewandten Krystallographie. Vol. II.

J. Weisbach: Anleitung zum axonometrischen Zeichnen. Freiberg, 1857.

C. Klein: Einleitung in die Krystallberechnung, pp. 381–393. Stuttgart, 1876.

E. S. Dana: Text-book of Mineralogy, Appendix B. New York, 1883. 2d Ed.

Th. Liebisch: Geometrische Krystallographie, Cap. IX. Leipzig, 1881.

V. Goldschmidt: Ueber Projection und graphische Krystallberechnung. Berlin, 1887.

INDEX.

Acanthite, 152, 207
Adjustment (goniometer), 23
Aggregates (crystal and crystalline), 16, 180
Albite, 199, 209
Amalgam, 61
Amorphous substances, 7
Andalusite, 153
Angle β, 159
Anorthite, 177
Apatite, 121
Aragonite, 188
Arsenopyrite, 196
Augite, 198
Axes, crystallographic, 24, 47, 82, 105
    of symmetry, 34
    twinning, 184
    zonal, 217
    projection of, 238
Axial planes, 24, 82, 105
    ratio, 83

Barium nitrate, 80
Basal edges, 87
    pinacoid, 90, 114
Beryl, 116, 233
Bevelment (of edges), 38
Boracite, 75
Boron, 95
Brachydiagonal axis, 143, 171
Brachydomes, 148, 174
Brachypinacoid, 149, 174
Brachyprisms, 148
Brachypyramids, 147, 172

Calamine, 43, 157, 187
Calcite, 130, 194, 195
Calcium hyposulphite, 175
Cane sugar, 169
Centering (goniometer), 23
Cerussite, 153, 196
Chalcocite, 151, 197
Chalcopyrite, 92, 102, 194

Chrysoberyl, 189
Clinodiagonal axis, 159
Clinodomes, 163
Clinopinacoid, 163
Clinoprisms, 163
Clinopyramids, 162
Clinographic projection, 237
Combination, crystal, 36
    oscillatory, 183, 209
Combinations, isometric, 59, 69, 74, 80
    tetragonal, 93, 99, 102
    hexagonal, 114, 121, 127, 136, 140
    orthorhombic, 150
    monoclinic, 164
    triclinic, 175
Composition face (twins), 185
Compound twins, 188
Congruent forms, 67 [20
Constancy of interfacial angles,
Contact twins, 185
Corrosion of crystal planes, 211
Copper, 192
Copper sulphate, 176
Crystal (def.), 1
    figures, construction of, 241
    form (def.), 35
        types of, 36
    growth, 10
    habit, 11
    series, 92
    systems, 43
Crystallizing force, 9
Crystallographic axes, 24
    constants, 144, 159, 171
    notation, 27
Crystallography (def.), 3
Cube, 56
Curvature of crystal planes, 210
Cyclic twins, 188

Dana's system of crystallographic notation, 29, 127

Diamond, 187, 210
Didodecahedron or Diploid, 67
Digonal axes (isometric), 47
Dihexagonal prism, 114
Dihexagonal pyramid, 109
Diopside, 166
Dioptase, 140
Disthene, 200
Distortion, 13, 205
Ditetragonal prism, 90
    pyramid, 87
Ditrigonal prism (hexagonal), 134
Dodecahedron, rhombic, 55, 177
    pentagonal, 68
    tetrahedral-pentagonal, 78
Dodecants, 25, 105
Dolomite, 210
Domes (def.), 36
    brachy, 148, 174
    clino, 163
    macro, 148, 174
    ortho, 163
Enantiomorphous forms, 66
Epidote, 167, 198, 202
Epsom salts. 156

False planes, 212
Fergusonite, 99
Fixed forms, 57
Fluorspar, 192, 206
Fundamental circle, 232
    form (= ground-form), 47, 83, 106

Galena, 61, 211
Garnet, 61
Gold, 206
Goniometer, 21
Gypsum, 186, 210
Gyroidal hemihedrism (isometric), 63

Habit of crystals, 11
Hematite, 130, 186
Hemihedrism (def.), 39
    gyroidal (isometric), 63
    pentagonal (isometric), 66
    tetrahedral (isometric), 71
    trapezohedral (tetragonal), 97
    pyramidal (tetragonal), 98

Hemihedrism, sphenoidal (tetragonal), 100
    trapezohedral (hexagonal), 118
    pyramidal (hexagonal), 119
    rhombohedral (hexagonal), 122
    sphenoidal (orthorhombic), 155
    monoclinic, 167
Hemihedrons (def.), 40
    apparently holohedral, 64, 68, 73
Hemimorphism (def.), 42
    tetragonal, 103
    hexagonal, 140
    orthorhombic, 156
    monoclinic, 168
Hemi-orthodomes, 163
Hemi-pyramids, 162
Hexagonal system, 45, 104
Hexahedron (= cube), 56
Hexoctahedron, 51
Hextetrahedron, 71
Holohedral, 40
Holohedron, 40
Hypersthene, 154

Icositetrahedron, 55
    pentagonal, 63
Ideal crystal forms, 14   [69
Inclined-face hemihedrism, 42,
Inclusions, 213
Indices, 26
    of a zone, 219
Individual crystals, 16
Integrity of zones, 221
Intermediate axes and planes of symmetry, 82, 106
Iodosuccinimide, 103
Iron vitriol, 165
Isometric system, 44, 46
Isomorphous growths, 181, 201

Juxtaposition twins, 185

Kernal crystals, 13

Lévy's system of crystallographic notation, 32
Limiting elements of crystals, 18
Limiting forms, 57, 76, 91, 114, 150, 164, 175

www.ingramcontent.com/pod-product-compliance
Lightning Source LLC
Chambersburg PA
CBHW021349230426
43666CB00006B/460